BUILDING WITH STRAW BALES

A practical guide for the UK and Ireland

Barbara Jones

GREEN BOOKS

Visual guide to the contents

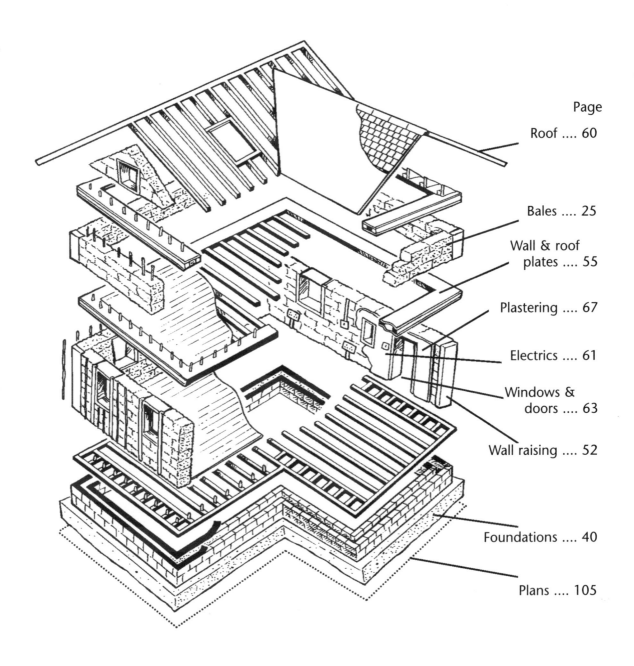

Contents

First published in 2002
by Green Books Ltd
Foxhole, Dartington, Totnes, Devon TQ9 6EB
<www.greenbooks.co.uk>

Reprinted 2003, 2005

All illustrations & esign
<deaftdesign

Techr

Design by Rick Lawrence <rick@see.globalnet.co.uk>

Main cover photograph © Barbara Jones

Printed by Biddles Ltd, King's Lynn, Norfolk, UK
on Norwood wood-free paper

Acknowledgements

The majority of this book was originally written with funding from the DETR under the Fast Track: Innovation in Construction Scheme. I would also like to acknowledge the invaluable input to this book by my partner Bee Rowan, without whose knowledge and attention to detail it would have been much less complete. Juliet Breese's illustrations are, as always, an enhancement of my work— thanks, Juliet; and Susan Hutchinson's work on CAD drawings is much appreciated.

Chapter 1

Introduction

Straw bale building is a smart way to build. It's more than just a wall building technique that has yet to come into its own. It's a radically different approach to the process of building itself. Like all innovative ideas, it has been pioneered by the passionate, and used experimentally by those with the vision to see its potential. Its background is grassroots self-build; it is firmly based in that sustainable, 'green building' culture that has brought to the construction industry many new and useful ideas about energy efficiency and responsibility towards the environment.

It is now at a pivotal point in its development, ready to be taken on by construction firms who see its value in terms of cost-effectiveness, sustainability, ease of installation and energy efficiency. As you will see from this guide, the building method itself is based on a block system, making the designs very easy to adapt from one project to another, and giving great flexibility in its use.

The accessible nature of straw as a construction material means that people unfamiliar with the building process can now participate in it. This opens the door for interest groups to work together on joint projects. Housing Associations and Local Authorities etc., are ideal managers for self-build straw projects that won't take years to complete, and which will engender an excitement and motivation that gets the job done. The atmosphere on a straw bale building site is qualitatively different to that found on the vast majority of other sites. It is woman-friendly, joyful, optimistic and highly motivated. Knowledge and skills are freely shared, and co-operation and teamwork predominate, all of which has a positive effect on health and safety on site.

Working with straw is unlike working with any other material. It is simple, flexible, imprecise and organic. It will challenge your preconceptions about the nature of building and the correct way of doing things; not everyone will be able to meet this challenge. Its simplicity can be disarming, or alarming. If you need complexity for

security, then this may not be for you. Don't be put off by nursery tales about the big bad wolf—we should be wise enough to realise that the wolf probably worked for the cement manufacturers! And don't pay too much attention to colloquial tales about 'hippie' houses—read on, and make your own mind up.

Straw as a building material excels in the areas of cost-effectiveness and energy efficiency. If used to replace the more traditional wall-building system of brick and block, it can present savings of over £4,000 on a normal 3-bedroomed house (see page 16 for more details). Of interest to the home owner is the huge reduction in heating costs once the house is occupied, due to the super insulation of the walls. Here the potential savings are up to 75% compared to a conventional modern house. Building regulations were revised in 2002, bringing the allowable U-value of domestic external walls down to 0.35 (the European Union would like to see 0.25), which is challenging the whole industry to meet these requirements. **A typical bale of straw has a U-value of 0.13—significantly better thermal performance than will be required.**

For more information on the U-value of straw, see page 84

This Guide is aimed at self-builders as well as the construction industry. It is meant to give clear and straightforward information about how to build houses with bales of straw. Since this is a simple and accessible wall building technique available to almost anyone, it is ideal for self-builders as well as mainstream builders at the forefront of sustainable house building. The language and descriptions are necessarily basic to ensure full understanding by everyone, particularly of first principles, and how and why we build with straw.

Throughout this book, I will be attempting to encourage you towards the best possible ways of doing things as far as current knowledge allows. It's always good to bear in mind though, that you are involved in a building process that is still developing—one which is simple, straightforward and based on common sense.

One of the biggest attributes of straw bale building is its capacity for creative fun, and its ability to allow you to design and build the sort of shape and space you'd really like. It lends itself very well to curved and circular shapes, and can provide deep window seats, alcoves and niches due to the thickness of the bales. It's also a very forgiving material, can be knocked back into shape fairly easily during wall-raising, doesn't require absolute precision, and can make rounded as well as angular corners. Partly due to its great insulation value and partly because of its organic nature, the inside of a straw bale house feels very different to a brick or stone one, having a cosy, warm quality to it and a pleasing look to the eye. The beauty of straw (apart from its aesthetic beauty) is that it combines very high insulation properties with great load-bearing potential: **a material that is building block and insulation all in one**.

Different styles and opinions have grown up around the world as bale building has spread. What was suitable in one climate has not proved to be best practice in others, and availability and cost of materials varies from country to country. However, there have been wonderfully imaginative adaptations to conditions. The main concerns in Ireland and the UK have been to do with:

- splash back—rain bouncing up from the ground onto the base of the walls

- rain causing high humidity in the surrounding air for long periods of time (only a potential problem at very cold temperatures)

- wind-driven rain.

Most of the differences in technique in this climate are to do with foundation design and the type of render used as a weatherproof coating. We have been able to draw on the rich knowledge of the past, using ideas which have been tried and tested over centuries. In many respects, the requirements of straw bale buildings are essentially the same as traditional cob (earth) buildings. They have high plinth walls, self-draining foundations, and large overhangs to the roof, "a good hat and a good pair of boots" as cob builders used to say. They are also constructed of breathable materials and must not be waterproofed (although they must be weatherproofed). There are currently over 100,000 cob houses of 200–500 years old still inhabited in the UK. For more information on cob houses, see the Reference Section on page 95.

Straw is a flexible material and requires us to work with it somewhat differently than if it were rigid. Accurate measurement and precision is impossible and unnecessary with straw, but working without these aids can be worrying to the novice, and threatening if you're already used to 20th century building

The beauty of straw is that it combines very high insulation properties with great load-bearing potential: a material that is building block and insulation all in one.

techniques. It is very important to make this clear at the outset. You have to develop a feel for the straw. You have to give it time, absorb its flexibility. Yet it *is* possible to be macho about it—to hurl bales around single-handedly and force them tightly into spaces, but this always has adverse consequences. Rushing the process, and working alone or competitively can mean that an adjoining section of wall is distorted and pushed out of shape—a section that someone else has spent time and care in getting right. It's as much a personal learning process as it is learning a new building technique. More than any other material (except perhaps cob and clay) it is susceptible to your own spirit and that of the team. Straw bale building is not something to do alone. It requires co-operation, skill-sharing and common sense. Many of the inspirational and artistic features occur in this atmosphere. It is empowering, expanding the world of opportunities for you and making possible what you thought to be impossible!

Building with bales can be inspiring and transformative, and working together with a group of people to build your own home can be one of the most empowering experiences of your life.

The atmosphere and environment in which we live is becoming increasingly a matter of concern to home owners and designers alike. There is a growing body of knowledge on the harmful effects of living long-term with modern materials that give off minute but significant amounts of toxins, the so-called 'sick-building syndrome'. Living in a straw house protects you from all that. It is a natural, breathable material that has no harmful effects. Hay fever sufferers are not affected by straw, as it does not contain pollens. Asthmatics too find a straw bale house a healthier environment to live in. Combined with a sensible choice of natural plasters and paints, it can positively enhance your quality of life.

When building a house using bales of straw, it's important to remember that it is the wall building material which is different. This has implications for the type of foundation required and can affect certain design decisions to do with windows, doors, roof bearing and render/plaster finishes. Otherwise, all aspects of the rest of the building remain the same. The installation of plumbing, electrics, interior carpentry, joinery and partition walls may be no different to the methods and materials you are used to. (Of course they could

Building with bales can be inspiring and transformative, and working together with a group of people to build your own home can be one of the most empowering experiences of your life.

also be re-thought in terms of using sustainable, locally sourced and recycled materials, but this is beyond the scope of this book). This Guide therefore covers details of different types of foundation, how to build walls with straw and stabilise them, how to protect walls from the weather and make them durable and how straw bale buildings can easily meet building regulations requirements. There is also a section on frequently asked questions, and a reference section for further reading, research and contacts.

History

Straw bale buildings were first constructed in the USA in the late 1800s, when baling machines were invented. The white settlers on the plains of Nebraska were growing grain crops in an area without stone or timber with which to build, and whilst waiting for timber to arrive by wagon train the following spring, they built temporary houses out of what was, to them, a waste material—the baled up straw-stalks of the grain crop. They built directly with the bales as if they were giant building blocks, where the bales themselves formed the loadbearing structure. This is known as the Nebraskan or loadbearing style. The settlers discovered that these bale houses kept them warm throughout the very cold winter yet cool during the hot summer, with the additional sound-proofing benefits of protection from the howling winds. Their positive experience of building and living in straw bale homes led to the building of permanent houses, some of which are still occupied dwellings today! This early building method flourished until about 1940, when a combination of war and the rise in the popularity and use of cement led to its virtual extinction. Then, in the late 1970s, Judy Knox and Matts Myhrman among other pioneers of the straw bale revival, rediscovered some of those early houses and set about refining the building method and passing on this knowledge to an eager audience of environmental enthusiasts. Through the green and permaculture movements the ideas spread very rapidly, with most of the new buildings being this self-build, Nebraska/ loadbearing style (see page 18 for more details). Before long, new

For more historical information, see The Straw Bale House by Steen, Bainbridge and Steen in the Reference Section, page 97.

techniques were developed to improve the building method and *The Last Straw* journal was founded in Arizona to disseminate ideas, promote good practice, and provide a forum within which owners and builders could network.

The first straw building in the UK was built in 1994, and the first in Ireland in 1996; and today approximately 1000 new structures are being built annually all over the world. There are about 70 in the UK and 10 in Ireland at the present time, some with full planning permission and building regulation approval. About twelve new buildings are constructed each year, including in 2002 the first phase of a residential school to be built using the loadbearing method.

Although the UK began building with straw bales earlier than any other European country except France, we have since fallen far behind in terms of official recognition and encouragement of this innovative and pioneering technique. There is an acute need for comprehensive research and testing of designs under different conditions, particularly under the sort of prolonged wet winters that we experience on our western coasts and uplands. Whilst empirical evidence is reassuring, there is still the need to know how these buildings will survive in the long term in our climate. It is also time for mainstream construction firms to realise the potential of straw, and to develop more commercial methods for building several houses at once.

Why use straw?

Sustainability

Straw is an annually renewable natural product, grown by photosynthesis, fuelled from the sun. Approximately 4 million tonnes are produced surplus to requirements each year in the UK. Using straw can mean less pressure to use other more environmentally damaging materials and in the unlikely event that the building is no longer required, it could be composted afterwards. (For further reading, see *Straw for Fuel, Feed & Fertiliser* in the Reference Section, page 97).

Energy efficiency and greenhouse gas emission

Over 50% of all greenhouse gases are produced by the construction industry and the transportation associated with it. If the 4 million tonnes of surplus straw in the UK was baled and used for local building, we could build at least 450,000 houses of 150m2 per year. That's almost half a million super-insulated homes, made with a material that takes carbon dioxide and makes it into oxygen during its life cycle. Coupled with vastly reduced heating requirements, thereby further reducing carbon dioxide emission (greenhouse gas) from the burning of fossil fuels, **straw bale building can actually cause a net decrease in greenhouse gas emissions**. To improve the energy efficiency of houses is increasingly becoming the design challenge of the 21st century.

To improve the energy efficiency of houses is increasingly becoming the design challenge of the 21st century.

Highly insulating

Straw provides super-insulation at an affordable cost. The K value of straw in a straw bale is **0.09W/mK** (see page 84); this combined with walls typically over 450mm thick gives a U-value of **0.13W/m²K**, two or three times lower (i.e. better) than contemporary materials, and much lower than current building regulations that require walls to have a U-value of 0.35 (see chapter on building regulations, page 81). When used with a design incorporating some thermal mass to store heat and release it on a 24-hour cycle (e.g. brick, clay, earth or cement), this maximises solar gain. This will dramatically reduce the amount of fuel needed to heat a straw bale house.

Sound Insulation

Straw bale walls are also super-insulative acoustically. There are two recording studios in the USA built of straw bales for their sound proofing quality and insulation. Straw bale wall systems are increasingly being used near airport runways and motorways in the USA and Europe as sound barriers. See back issues of *The Last Straw* for details.

Low Fire Risk

Plastered straw bale walls are less of a fire risk than traditional timber-framed walls. Research in the USA and Mexico has shown that "a straw bale infill wall assembly is a far greater fire resistive assembly than a wood frame wall assembly using the same finishes." See section on Fire on page 86.

Low Cost

Straw is currently produced surplus to requirements. It is generally regarded as a waste product, and a bale costs on average £1.50 delivered or 40p from the field. The walls of a 3 bedroomed, two-storey house can be built with 520 bales, which cost £780 compared with a material cost of £5,000 for a brick and block wall. Also, because the building method is so straightforward, people without previous building experience can participate in the design and construction, thereby saving on labour costs.

The most significant saving on straw bale houses is in the long-term fuel reductions due to the high level of insulation. Heating costs can be reduced by up to 75% annually compared with modern style housing, and the savings therefore continue to accrue throughout the life of the building.

Structurally sound

Bales of straw are more than adequate to carry typical loadings of floors, roofs and winter snow loads. They have passed load-bearing tests both in the laboratory and empirically, and are used to build at least 2-storey houses. (See Reference Section for report on house in Nova Scotia, page 103.) When using a framework method, where the bales simply infill the gaps between posts, the bales do not take weight (although they could!) and it is the posts which carry the load.

A Healthy Living Environment

Straw, particularly organic straw, is a healthy alternative to modern materials. It is natural, and harmless. It does not cause hay fever since it's not hay, and in fact is the material of choice for many allergy sufferers because it is so innocuous. Living within straw walls can enhance the quality of air we breathe, because it does not give off harmful fumes such as formaldehydes, as many modern materials do, and because it is a breathable material, thereby helping to keep the inside air fresh. Coupled with the use of nontoxic organic finishes such as clay and natural pigments and paints, and with opening windows, it can provide one of the safest and most comfortable atmospheres in which to live. Another health benefit is the ambience inside a straw bale house which is calm, cosy and peaceful. This is partly to do with the high level of sound insulation, partly to do with the air quality, and partly to do with the organic feel to the house—a beautiful, nurturing and safe environment to inhabit. Try it!

Empowering and Fun!

The most unquantifiable aspect of a straw bale house has to be the way that the building process itself empowers ordinary people. It is accessible to many people who are otherwise excluded from the design and build process, and enables them to transform their living environment, and their lives, in a very enjoyable way.

Chapter 2

Different Methods of Building

*Loadbearing • Lightweight frame and loadbearing
• Infill and timber frame • Hybrid design*

Nebraska *(also called Loadbearing)*

This is the original method of building, pioneered by the Nebraskan settlers in the USA. In this method, the bales themselves take the weight of the roof; there is no other structural framework. They are placed together like giant building blocks, pinned to the foundations and to each other with coppiced hazel, and have a wooden roof plate on top, which spreads the floor and roof loads across the width of the wall. The roof plate is fastened to the foundations and the bales with coppiced* hazel and strapping, and the roof is constructed in the usual manner on top of the roof plate.

Windows and doors are placed inside structural box frames, which are pinned into the bales as the walls go up. This is the simplest method and the most fun way of building; it requires little previous knowledge of wall construction and is very accessible. Owner builders tend to prefer this method because of its simplicity, ease of design, minimal use of timber, and the opportunity it affords for a modern day wall raising. The potential for empowerment through working together on a shared project is one of the main differences between this type of building and any other.

---

*Coppicing is the ancient traditional craft of managing mixed woodland trees, harvesting different varieties for specific uses such as furniture making, hurdle making, shipbuilding (not so common now!), supplying the steel industry with fuel, basket making etc. The trees are cared for and cut in rotation, and new growth encouraged for the next harvest.

Advantages:

- A simple, straightforward and accessible building method

- Easy for non-professionals to design, following readily comprehensible basic principles.

- Designs from one room to two-storey homes can be created using a simple, step by step approach.

- Curves and circles are easy to achieve, for little extra cost.

- Ideal for self-builders because of its simplicity, accessibility, ease of design, and low cost.

- The straw is very forgiving. Total accuracy in plumb is not a design goal but wilder variations can be brought back into shape easily!

- Great versatility of design shape.

- **It's fast!**

Disadvantages:

- The straw must be kept dry throughout the whole building process until it is plastered, which can be very difficult on a large building, or one that is being constructed slowly.

- Openings for windows and doors should not exceed 50% of the wall surface area in any wall.

The Nebraska style is the most common method of building to be found in Ireland and the UK. However, for larger buildings, the following method is also appropriate.

Lightweight Frame and Loadbearing

This design is a way to retain the benefits of the loadbearing style, yet enabling the roof to be constructed before the straw walls are built, thus giving protection against the weather throughout the wall-raising process. It uses a timber framework which is so lightweight that it could not stand up alone, and which requires temporary bracing and props to give it stability until the straw is in place. The straw is an essential part of the structural integrity of the building, *more so than the timber,* and it works together with the timber to carry the load of floors and roof. Timber posts are located at corners and either side of window and door openings *only*, and are designed such that the timber wallplate at first floor and/or roof level can be slotted down into the posts once the straw is in place, allowing for compression on the bales (see diagram on page 56). Compression of the straw bale infill walls is *essential* for stability. To increase stability, the bales are pinned externally, and the pins are secured onto the base and wall plate of the framework once all the settlement of the walls is complete. It is constructed in such a way that the wallplate and roof are held 100mm above the finished straw wall height whilst the wall is being built. Once the bracing and props are removed, the roof is lowered into the slotted posts, which causes compression of the straw beneath it. The roof weight can be encouraged to compress the walls faster by strapping it down to the foundations and mechanically compressing the walls using ratchet straps.

> *One of the most important design features of a loadbearing straw bale house is to distribute the loads as evenly as possible around the whole building. Never use point loads.*

Advantages:

- The roof can be constructed before the straw is placed, giving secure weather protection.

- Framework and posts can be constructed off site.

- Provides greater stability for window and door frames than in the loadbearing style.

- Vastly reduces the amount of timber required compared to the more traditional post and beam method.

Disadvantages:

- It is more complicated than the Nebraskan style to construct.

- Greater technical ability is required to make the structure stable whilst the straw is being placed.

Infill (also called Post and Beam or Timber Frame)

In this method, the weight of the roof is carried by a wood, steel, or concrete framework; the bales are simply infill insulation blocks between the posts. This has often been the preferred option for architects, as the structural concepts are not innovative and rely on an already established method of building, therefore the risk associated with an experimental technique is minimized. There is no need to satisfy oneself of the capacity of the bales to take the roof weight, since the framework does this. This method requires a high level of carpentry skill and uses substantially more timber than a loadbearing design, so has significant cost and environmental implications.

There are many types of straw buildings that use a combination of ideas from the above techniques, or use new ideas. It is still an experimental method, and being so simple, allows for invention during practice.

Advantages:

- The roof can be constructed before the straw is placed, giving secure weather protection.

- Framework and posts can be constructed off site.

- Provides greater stability for window frames than in the loadbearing style.

- In conjunction with a steel frame, can create large open spaces such as warehouses or industrial units.

Disadvantages:

- It is more complicated than the Nebraskan style to construct.

- It requires a high level of carpentry skill (or metalwork experience in the case of a steel frame) to construct the frames.

- It uses a large amount of timber.

Hybrid Methods

There are many types of straw buildings that use a combination of ideas from the above techniques, or use new ideas. It is still an experimental method, and being so simple, allows for invention during practice. For instance, it's possible to build well-insulated loadbearing walls to protect your house on the cold North side and combine this with a framework method on the South, allowing for lots of windows to maximize solar gain.

Mortared Bale Matrix

This method was pioneered in Canada twenty years ago by Louis Gagné. Here the bales are used much more like bricks, with cement mortar holding them all together. The bales are stacked in vertical columns so the cement mortar, in effect, forms posts between each stack. The whole building is cement rendered inside and out. It is rarely used now because of the knowledge of simpler, and more enjoyable methods.

Advantages:

- It is very effective, and has been thoroughly tested in Canada.

Disadvantages:

- It is very labour-intensive.

- It uses a lot of cement.

- It is susceptible to damp caused by the use of cement render on straw.

- It falls into the category of 'no fun' building methods.

Raise the first course of bales up from the ground by at least 225mm (9"), put a 450mm (18") overhang on the roof to protect the walls from rain, and you won't go far wrong.

Other Aspects of Straw Bale Building

The methods described above are for a type of wall building system that is different to the methods we have become familiar with in the 20th century. All other aspects of the building remain the same, including plumbing, electrics, roofing etc. The main differences, as mentioned above, would be found in the design of foundations, type of wall building material, and type of render or plaster.

Straw, being a breathable material, functions best when used with other such materials. Therefore, it is common to design foundations without using cement, or where cement is used, to protect the straw from it by using a different material in between, usually timber plus a damp proof course, and to incorporate drainage into the foundation itself. In the same manner, cement renders and gypsum plasters would not be used, but instead, traditional lime, and/or natural clay renders and plasters would be applied. Most straw bale houses, of whatever type of construction, are rendered inside and out, so that when finished they can look very similar to a traditional style cottage, very beautiful and with deep walls—it is hard to tell that they are made of straw. Several coats of lime-wash are essential as a surface finish and weatherproofer, and this must be re-applied, as with all other painted houses, every few years.

The key to durability with a straw bale house, as with any other, lies in good design and detailing, quality work, & maintenance when necessary throughout its life.

Durability

Because of its simplicity, it is possible to build a wide range of different quality structures, from a straw bale shed to last 10 years, to a straw bale house to last upwards of 100. Straw bale building is still a relatively new concept, and as such some areas of design are still experimental. In the UK, the oldest buildings have stood for only 7 years, and some of the early ones were never intended to be more than experiments. However, there are now under construction homes for families, classrooms and whole schools, centres for community groups as well as numerous owner-built houses, offices, studios and garages, animal shelters, food and machinery storage barns, etc.

No straw bale building in the UK or Ireland has ever been refused planning permission or building regulation approval on the grounds of it being made of straw, or on the question of durability.

Design Pointers

- Raise the first course of bales up from the ground by at least 225mm (9"), put a 450mm (18") overhang on the roof to protect the walls from rain, and you won't go far wrong.

- One of the most important design features of a loadbearing straw bale house is to distribute the loads as evenly as possible around the whole building. Never use point loads.

- The key to durability with a straw bale house, as with any other, lies in good design and detailing, quality work, & maintenance when necessary throughout its life.

- Avoid using metal in the walls if at all possible, since it is a cold material and may encourage warm, moisture-laden air from the inside of the house to condense on it.

- Loadbearing houses are subject to settlement as the straw compresses under the weight of the floors and roof. Allowance for this must be designed in by leaving settlement gaps above doors and windows. See section on windows, page 63.

Chapter 3

Bale Specifications

The Nature of Straw

If we leave a bale of straw out in the field to be rained on, it quickly becomes too heavy to lift because of water saturation and is of no use other than as mulch for trees. However, if we stack lots of bales carefully out in a field, raise them off the ground and put a good roof over the top, they will withstand the weather and the outside edges simply get wet and dry out. Talk to any older farmers and they will tell you this is how straw (and hay) was traditionally stored—in the field for ease of access. They would raise the bales off the ground first, usually by using a sacrificial layer of bales laid on edge (i.e. one that would go to waste later), and the rest stacked flat, with a roof of thatch over the top. The sides of the bales would be exposed to the rain and wind, but getting wet was not a problem. Straw does not 'wick' (suck) water into itself like concrete does. It simply gets wet as far as the force of the wind can drive the rain into it. When the rain stops, the natural movement of air or wind around the bales dries them out. This cycle of wetting and drying does not damage the bale.

It is important not to let the centre of the bales get wet through the top or bottom, as they are unlikely to dry out sufficiently for building, but wetting the sides of a bale is not usually a problem.

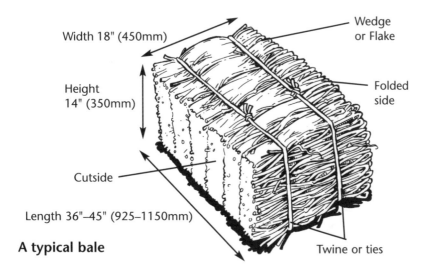

Width 18" (450mm)

Wedge or Flake

Height 14" (350mm)

Folded side

Cutside

Length 36"–45" (925–1150mm)

A typical bale

Twine or ties

How to Choose Good Building Bales

Bales should be dry, well compacted with tight strings, be of a uniform size and contain virtually no seed heads. They must not be damp, and must be protected from damp during the building process. Safe moisture levels for the prevention of fungal and bacterial growth are as follows:

Either: **moisture content should not exceed 15%.** This means that the weight of water in the bale should not be more than 15% of the weight of the same bale if it was thoroughly dried.

Or: **relative humidity should not exceed 70%.** This means that whatever the maximum amount of water vapour the air can hold before that vapour condenses out as water, the bales must not contain more than 70% of it. The relative humidity of air changes with temperature. Air can hold more vapour if it is warmer.

- Bales should be as dense and compact as possible. The baling machine should be set to maximum compression; in general this means bales contain about one third more straw than usual. Weight should be between 16–30 kg.

- Bales should be roughly twice as long as they are wide, and the larger the better. Most baling machines produce two-string bales that are 18" (450mm) wide x 14" (350mm) high and variable lengths from 36"–45" (900–1125mm), although a few machines are 20" (500mm) wide and 15" (390mm) high. Mini-heston bales of 8ft x 3ft x 2ft can also be used, and are especially good for building extremely large spaces such as warehouses.

- Strings must be very tight, so that it is difficult to get your fingers underneath. They should be about 100mm (4") in from the edges of the bale and not sliding off the corners. String should be polypropylene baling twine, sisal or hemp, not wire.

- The type of straw is immaterial as long as the above guidelines are followed. It can be wheat, barley, rye, oats, etc. Straws should be long, 150mm (6") min, preferably 300–450mm (12–18").

Do not confuse straw with hay or grasses. Straw is the baled-up dead plant stems of a grain crop. It has had virtually all its seed heads removed, and contains no leaves or flowers. It is a fairly inert material, with a similar chemical make-up to wood. It is quite difficult to make it decompose, and usually requires the addition of nitrates to do so. Hay, on the other hand, is grass baled up green, with lots of feedstuff (leaves and flowers etc.) deliberately left in there. It readily decomposes, as the organic matter in it begins to rot.

The age of the straw does not matter as long as the above conditions have been followed, and it has been stored correctly. All the above conditions should ideally apply equally to all bales, whether they are being used for loadbearing or infill.

It is important to know the size of bales you will be using before finalising dimensions of foundation, wallplate etc.

Bales can vary a lot in length, from supplier to supplier and within each load, as it depends on the skill of the tractor driver and the evenness of the field as to whether or not the straw is picked up uniformly as it is baled.

In practice, relying on the farmer to tell you the length of bales is not a good option. Besides which, you will need to satisfy yourself that the straw is baled dry, and kept dry whilst in transit and storage. Far better to look at the bales once they're harvested and determine the average length of bale at the same time. The best way of doing this is to lay 10 bales tightly, end to end. Measure the whole and divide by 10 to find the average. (It is almost impossible to measure the length of one bale accurately). However, if you find that your delivered bales are not the same length as you expected, it's not an insurmountable problem. It may mean a little more work in shaping the bales to fit, but this is straightforward and not too time-consuming.

As straw bale building becomes more widespread, suitable construction bales will become more readily available. Already there are wholesalers in the UK who can supply good construction bales. These can be found under Agricultural Merchants in the Yellow Pages or Thomsons.

It is possible to harvest and store straw in bales of uniform length and moisture content, ready for the building market each season. As demand increases, so will the reliability and availability of supply.

The Cost of Bales

The cheapest way to buy bales is straight off the field after they've been made, and to buy locally so as not to pay large transport costs. This has the added benefit of minimising the environmental impact of transportation. If you collect them yourself they can cost as little as 40p per bale.

When you consider the average 200m² 3-bedroomed house will use about 520 bales, this represents a material cost of only £208!

Even when buying in bulk from a wholesaler, delivered to the site, bales will cost on average £1.50 each, which would bring the price of 520 bales up to £780. (Of course, this cost is likely to increase as the demand for straw bales as a construction material increases!)

Compare this to the material cost of building the same walls in brick and block:

520 bales of size 1.1m x 0.35m = 200m² approx

200m² of brick and block walling uses 12,000 bricks and 2,000 blocks. If bricks cost 30p each and blocks 70p, the total cost would be £5000.

So the first financial saving in building such as a typical 200m², 2-storey, 3 bedroomed house with straw instead of brick and block is approximately:

£5,000 – £780 = £4,220

On top of these costs, you also have labour costs to calculate, with brick and block walls taking a team of 2 skilled people and 2 labourers an average of 6 weeks to complete, and straw bale walls taking a team of 10 unskilled volunteers plus a trainer about 2 weeks to complete.

This level of savings from using straw as a loadbearing material instead of brick and block increases with the size of the building.

Although the walls of a building only represent about 16% of the total costs of a finished building, £4,220 is a significant saving for any self-builder and becomes more so for construction firms building more than one house. Together with this, the labour times involved in straw bale building are vastly reduced once the labour force is familiar with the material, and the role of trainer becomes redundant.

In addition, this affects the design of the foundation which can use less material because straw weighs on average 65% less than brick, and has a wider bearing surface, so spreading the loads further.

Can you afford not to build with straw?

Chapter 4

Bale Plans

Getting Started

Think about what you want your straw bale house to look like and how you want it to feel inside. Try to forget anything you've been told about building and imagine your ideal space, however wild that might seem! Then work within the practical limitations of the bales to come as close as possible to your dream.

The design of straw bale houses is usually simple and elegant. It is based on a block design and therefore different elements of the structure can be built up easily from the initial shape and dimensions of the foundations. Each section of the house has an obvious relationship to the other sections, and many different houses can be designed quickly and easily from the same basic plan.

For most domestic dwellings, it should be possible for owner-builders to design their own houses. The way a straw bale house goes together is simple. It follows common sense principles and it is effective. By using this guide, you should have no difficulty in working out the construction drawings and methods for any type of domestic dwelling.

Once you've decided on what the building is for, what you want it to look like, and what you want it to feel like, begin by reading the section on the nature of straw (page 25). Think about its orientation, try and make it face south, and have more or larger windows in this side to maximise solar gain.

Draw the shape of the building you require, as though you were looking at it from above, this is called the PLAN view. Draw in the shape of the bales, their width and length, planning where they lie on the first course of the wall, as in the drawing opposite.

Now imagine you are looking at the finished building, standing on the ground looking north, south, east and west. Draw the face of the building you see from each direction, showing again where

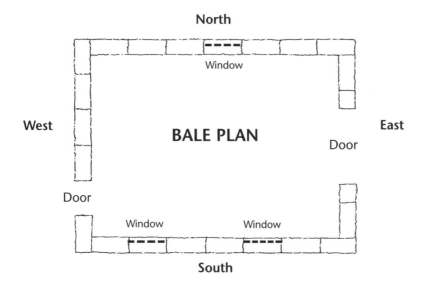

North

Window

West

BALE PLAN

East

Door

Door

Window   Window

South

each bale is and how they turn corners or curve etc. These drawings are called ELEVATIONS (see page 32).

From accurate bale plans you can work out how many bales you need, how many hazel pins, (2 per bale from the 4th course up) staples (in every bale where it changes direction) and other quantities of materials. Details of foundations, windows, first floor and roof can be worked out. You also now have the basis for drawing your own plans for planning permission.

Principles of Bale Design and Layout

- Before you draw your final plan, and before you finalise the dimensions and lay out the foundations, you need to know the dimensions of the bales you will be using as they can vary a lot!

- The bale plan should be made up of a whole number of bales.

- Do not have any places in the wall (e.g. beside a window) that are less than half a bale length.

- Window and door openings must be at least one bale length away from corners in load-bearing designs.

You will need to start by planning where the bales lie on the first course of the wall.

- If at all possible, choose window and door sizes that are multiples of bale dimensions.

- try to design frameworks so that the distance between posts is equal to a whole number of bales or half bales, reducing the labour time involved in customising.

In a loadbearing design the walls will settle a bit once the weight of the roof is on, so allow for this by leaving gaps above windows and doors that can be filled in later. With good building bales, settlement in a seven bale high wall should be about 12–50mm (½"-2"). The amount of settlement depends on the density of the bales and the amount of loading applied to them (such as the weight of the roof, if there is more than one floor, etc).

Loadbearing houses are subject to settlement as the straw compresses under the weight of the floors and roof. Allowance for this must be designed in by leaving settlement gaps above doors and windows.

BALE ELEVATIONS

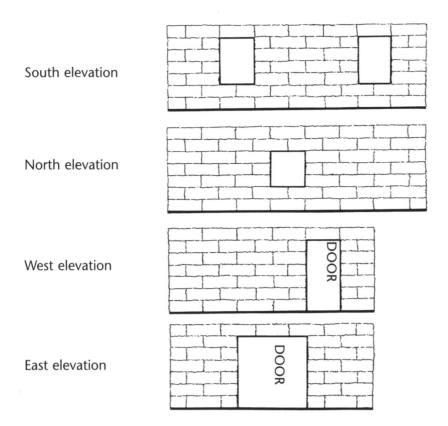

South elevation

North elevation

West elevation

East elevation

Chapter 5

Safety and Tools

In general, there will be fewer accidents and mistakes on site if everybody is happy and well motivated. It is good practice to encourage the whole team to be aware of each other, of each other's roles, and of how each separate action is part of the whole achievement. It can be very useful for everyone to meet together at the start of each day, to share some thought or reflection on the work to be done or the manner of its doing, and discuss possible problems or areas that will need co-operation or greater care. If we can recognise our interdependence with each other rather than be in competition, this will help to engender a caring and considerate workforce, where each individual is responsible for their own selves but also has regard for others' welfare.

This is especially necessary on a self-build site where people have different levels of skill or knowledge.

Straw bale building sites have become renowned for their ethos of working together, sharing knowledge and skills, equality for women, enjoyable learning and fun. But there is no reason why these qualities cannot be encouraged on *all* building sites.

Knife

Tape measure

Baling twine

Safety on Site

These are basic guidelines that will help to ensure that nobody has an accident:

- Every building site, even if it's your own home, should have a first aid kit available **and everyone should know where it is!**

- There should be no smoking anywhere on site or around stored straw.

- Always keep the site tidy. Scaffolding and working areas should be cleared and tidied each evening before finishing. It is helpful to store materials and rubbish etc. in specific places to keep a sense of order.

- Have a central place for tools so that any that are not being used can be put back, and can be found when needed.

- Never leave tools lying around. Each person should be responsible for the tools that they are using, should know where they are, and put them away when not in use.

- Sharp tools should be sheathed or put away when not in use.

- Unplug all electrical tools when not in use.

- Do not leave electrical leads trailing across the site.

- All ladders must be securely tied.

- Never move a ladder temporarily and then leave it unattended.

- Stepladders and scaffolding towers etc. should always have a firm footing.

- Hard hats should always be worn when anyone is working overhead.

- Extra care should be taken when using heat or open flames, e.g. when connecting plumbing pipes together.

Hay knife

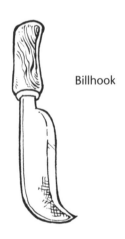

Billhook

Tools for Straw Bale Building

Straw bale building requires almost no specialist tools. A basic tool kit of items can often be found around the home, but if you don't already have them, they are relatively cheap to buy and not too difficult to find. As you become familiar with using tools, you'll find that particular ones suit you better than others, and you'll develop favourites. This may be because the handle is just the right size, or that it does its job well, or it's well balanced, or because you like the look of it and it's familiar. **Developing a feel for tools can be one of the unexpected joys of working with straw.**

As with most things, there are different standards of quality available on the market. On the whole, it is best to buy better quality tools even if they cost a bit more because **poor quality tools may be inaccurate, wear out quickly, or stretch!** They will certainly frustrate you and make you feel inadequate.

Some of the reasons for investing in good tools are:

• You will be using them over a long time period, so they need to last.

• They must withstand normal wear and tear on a building site, so need to be robust.

• Some will probably get lost in the straw at times, and should be able to withstand this, at least over night.

• Using tools that don't do their job well can make you think it's you not they that's the problem.

• Good tools, if well maintained, can last a lifetime, not just whilst you build your first house.

On the following two pages is a list of some of the most common tools you will need.

Baling needles

Persuader

Tool	Preferred type	Comments
Tape measure	5m long, with metric and imperial units together	The hook on poor quality tape measures may come loose easily, bend or break, or not be set correctly. They may also stop returning into their holder, rust easily, or snap!
Stanley knife	With retractable blade for safety	Any small hand knife will do, preferably on a string to avoid losing it!
Claw hammer	14oz weight or less, unless you are used to them	Cheap hammers bend when pulling nails out, or become pock marked on hitting nails. Their handles may pull off too!
Lump hammer	2lb is sufficient unless you're trying to impress someone.	The head can fly off cheap ones!
Hand saw	Cross-cut panel saw	A sharp saw is a dream to use, but a blunt one may put you off carpentry for life . . .
Axe	Side axe is best, but splitting axe is useful too	Get one that fits your hand size and strength.
Bill hook	Any type will do	This is an alternative to the axe, and extremely handy if you're used to using one.
Persuader	Long handle, large, lightweight head	Otherwise known as a fencing hammer. No straw baler should be without one!
Baling needles	Lightweight steel, pointed with 2 holes at one end	Should be long enough to go through the width of a bale, and sturdy enough not to bend too much.
Gloves	Fabric with criss-cross plastic	Most work gloves are made for men with big hands, so choose carefully . . .
Hedge clippers	Any type will do	Need to be really sharp, as straw is particularly tough to clip. Electric ones are not recommended as they clog up quickly.

Axe

Other Useful tools   Preferred type / comments

Combination square	For marking 90° and 45° angles	Cheap ones may bend or not measure angles accurately.
Bevel	For marking angles from 0° to 180°	Most common problems: the tightening screw handle getting in the way when marking, or not staying tight.
Hay knife	Large single blade	A traditional tool that is extremely effective, but may be difficult to find. Can be made by a blacksmith from an original.
Electric drill	Powerful, with hammer action	See 'power tools' below.
Earth trip	Plugs directly into a socket	An essential protection when using 240v power tools on site.
Screwdrivers	A selection of sizes, both straight head and posidrive or phillips	Can also use battery power. The heads on cheaper ones are made of softer metals that will wear out rather than turn a difficult screw.
Spanners and wrenches	A selection of sizes	May need both metric and imperial sizes.
Crowbar	Long but not too heavy	Can be used for many different jobs.
Sledgehammer	Short and/or long handled	Possible alternative to the Persuader, but can encourage too much macho behaviour.
Garden strimmer	Best with a metal cutting blade	All types will work, be careful not to get one that's too heavy, as it has to be used sideways.

Clothing and Safety Equipment

- When working with straw it is advisable to wear a long sleeved shirt and long trousers because straw stalks can be quite prickly.

- Gloves are also a must, at least for some aspects of the work.

- Overalls can be handy when doing particularly messy jobs such as plastering or painting.

- Goggles, ear-protectors and dust masks may be needed for certain jobs.

- Hard hats must be worn if anyone is working overhead

Hedge clippers

- Stout boots are needed to protect your feet on uneven surfaces and from falling objects

- As loose straw is a fire hazard, have a long enough hose pipe ready in case.

- Have a first-aid kit on site.

Power Tools and Electricity on Site

Electricity, at the voltage that it is supplied to us domestically, is powerful enough to kill outright if we make direct contact with it. This is why there are so many safeguards in the home, to ensure that this never happens. On major construction sites, the risk of cutting through a cable is so much greater than at home that power tools are usually used at a lower voltage than the normal 240 volts, so that any potential shock is not life threatening. Tools such as drills are connected via a yellow round pin socket to a transformer, which itself is connected to the mains electricity by a short lead. The transformer converts the usual supply of 240 volts to 110 volts. When using power tools at home, or on a self-build site, it is not common to use 110v equipment, and in fact 240v equipment is sold as standard in all DIY shops. When used according to manufacturers instructions this is not dangerous. However, on any building site, risks are much greater than when

Electrical tools from a hire shop will be supplied to work at 110v with a transformer.

Earth trip

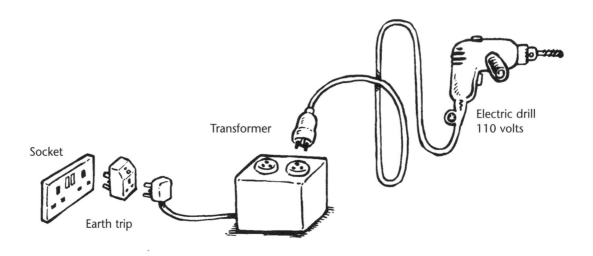

Socket

Earth trip

Transformer

Electric drill
110 volts

working at home, and it is a sensible precaution to use an earth trip plug for each electric tool. This looks like an ordinary adaptor that plugs into the socket; the electrical tool plugs into this. What it does is to cut off the power supply if there is a short in the circuit, such as would occur if the cable were cut or immersed in water. Electrical tools from a hire shop will be supplied to work at 110v with a transformer.

Battery-powered tools with rechargeable batteries are increasingly popular and useful. The main ones in use are drills and screwdrivers, but you can also buy circular saws, jigsaws etc. They have a huge advantage over standard plug-in tools in that they do not have any trailing leads to trip over or cut through.

Main things to consider are:

Lump hammer

- The tool must be powerful enough for the job it is required to do; otherwise it can be very frustrating.

- Batteries should store enough charge to give a reasonable length of time for usage.

- Two batteries are best, then one can charge whilst the other is being used.

Handy Tips

- If you buy brightly coloured tools, and/or mark them with coloured tape, they will be easier to find when you have inevitably lost them somewhere in the straw!

- Cut the strings on a customised bale at the knot, and hang them up in a convenient place so they can be re-used for the next customisation.

- Never stick baling needles into the ground as the holes get clogged with soil. For safety, and so you can find them again, stick them upright in a spare bale.

Chapter 6

Foundations

An Introduction

All buildings need to have some sort of a foundation on which to build. This may simply be the natural foundation of the earth beneath which may be bedrock, firm clay, compacted gravel etc., but we are more familiar in the 20th century with artificial foundations such as poured concrete strips and slabs. As the foundation has to carry the weight of the walls above it, and other loadings such as floors, furniture, roofs and even snow in winter, it is important to know what type of earth (or subsoil) is found on your building site. Different types of earth will carry different weights. Bedrock, for instance, will carry much greater weight than soft clay. On the other hand, if you increase the surface area that bears the weight onto soft clay—much like wearing snow-shoes in the snow—even this can take the weight of a house. For a small building constructed of lightweight materials, there is obviously no need to build massive artificial foundations on any type of soil. Equally, for a heavy building built on bedrock, there is no need to add huge foundations. Almost all the buildings in the UK and Ireland that are more than 200 years old have natural foundations with little or no artificial ones. They often use larger stones at the base of the wall, making it slightly wider than the wall itself. In all cases, they removed the topsoil (growing part of the earth) and dug down to something solid. Because they chose their building sites well, this was often only a few inches below the ground surface. There are hundreds of thousands of houses still lived in today, that can be excavated by only 6 inches or so to find they are sitting on the earth itself, and yet are completely sound and safe. Unfortunately, there are many misconceptions about foundations today that are partly caused by the rise in popularity of cement and concrete. In some building colleges, students are taught that buildings *must* have foundations made of concrete, despite the evidence to the contrary that surrounds us. Throughout learning about straw bale building, you will be encouraged to look at what's around you, to keep things as simple and straightforward as they can be. There is no need to overcomplicate anything, only to **understand what it is that we want to achieve and make choices based on the different ways that it is possible to do so.**

So for foundations, we want to achieve a solid, stable base that distributes the weight of whatever is built upon it over the ground beneath. We also want to be sure that there is no unequal settlement throughout the building.

If we look at the different weights of materials, we can see that using straw for the walls can have a significant impact on the choice, and cost, of the foundation.

For comparison:

$$\begin{aligned}
\text{1m}^2 \text{ of brick} &= 212\text{kg} \\
\text{1m}^2 \text{ of block} &= 197\text{kg} \\
\text{1m}^2 \text{ of straw} &= 75\text{kg}
\end{aligned}$$

Therefore straw weighs 65% less than brick and 62% less than concrete block. And when you consider that most modern houses have a standard cavity wall construction, one wall of brick and one of block, this further reduces the comparative weight of the straw wall to only 19% of a standard cavity wall.

A single storey structure, built with loadbearing straw walls, should not need more than a base plate the width of the walls to give it a secure foundation. That is, no need for deep trenches filled with concrete, perhaps no need for concrete at all.

Foundation Types Specific to Straw Bale Buildings in this Climate

Having understood the aim of natural and artificial foundations to provide a solid and stable base from which to build your house, we also need to pay attention to the specific requirements of the wall building material we are using, namely straw.

The base of a straw bale needs to be kept dry in the walls of a building. This means:

- It must be raised off the ground sufficiently to avoid damage by splashback from rain bouncing off the ground.

- There must be no possibility of moisture being trapped at the base of the straw, at the interface between straw and foundation.

Both of these can best be achieved by using **self-draining foundations**, ones in which any moisture that enters the foundation is encouraged to travel downwards and out of the building, rather than to wick upwards or sit where it is. But there are other reasons for using them too.

Why use self-draining foundations?

- They have withstood the test of time, and are a tried and tested method. Some of the oldest traditional buildings in the UK and Ireland, many over 400 years old, are made of cob (earth) and use self-draining foundations. There are significant similarities in the properties of straw bale and cob buildings and we can use the knowledge of generations to inform our practice today.

- It is sensible in the often wet and windy climate of the UK and Ireland, to use this type of foundation as an important protection against the severity of the weather. If moisture enters the bale walls, it will slowly migrate downwards into the bottom bale, where it will stay and damage the wall if the foundation doesn't drain.

- When the foundation is built up above ground level, it not only provides drainage for the wall, but also provides protection against the possibility of damp rising up through the wall from the earth beneath.
- Many people are trying to reduce the amount of cement they use in building (for environmental reasons) and a self-draining rubble trench instead of concrete is an option.
- Depending on the design, self-draining foundations can be built cheaply and without the need for professional builders.

Other Differences in Foundations Due to the Use of Straw

Tie-downs

Foundation design must incorporate some method that allows for the wallplate and roof to be fastened down securely to it. This prevents the roof from being lifted off by strong winds. It can be done in several ways:

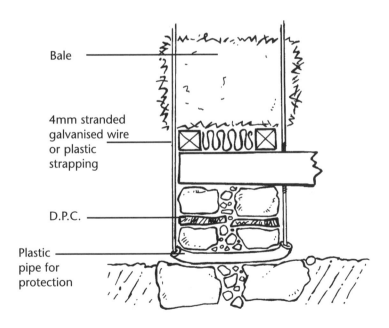

Bale

4mm stranded
galvanised wire
or plastic
strapping

D.P.C.

Plastic
pipe for
protection

FOUNDATION SHOWING ONE TYPE OF TIE-DOWN DETAIL

- Metal or plastic strapping can be laid underneath the foundations in a U-shaped plastic pipe for protection: ordinary garden hose pipe is sufficient. The strapping can then be carried over the wallplate once the straw is in position, and fastened in tension using fencing connectors or similar.

- Anchor bolts can be fixed into the foundations internally and externally to take metal or plastic strapping. Ensure they are fixed close enough to the wall so that they can be plastered in.

- Strapping can be fixed to the timber base-plate that is laid on top of the foundations.

Fixings for door frames etc

Anything, such as doorframes, that fix directly to the foundation, must have provision made for them. Structural box frames on concrete or stone are usually bolted into the foundation. They can also be fixed to the timber base-plate.

Different Types of Foundation

Local Stone with Timber Floor Grid

Advantages:

This is the most ideal type to choose, because:

- It is made entirely of natural materials

- The stone can be second-hand as well as new

- It is very beautiful to look at

- It is easy to build even with no previous knowledge

- It can all be re-used if ever dismantled

The plinth wall is built at least 9 inches high to protect the base of the straw from splashback, (the rain bouncing up from a hard surface onto the wall).

It is not necessary to build the plinth on top of a draining trench if the ground below you is sufficiently stable to support the weight of the building, e.g. stone, gravel, or compacted clay. You may want to use a shallow rubble trench if the ground does not drain well.

Disadvantages:

- If you are not doing it yourself, labour costs for stone building are much greater than for concrete block or other types of foundation.

- If the stone is not second-hand, or found, it is expensive compared to concrete block or other types of foundation.

- It is a slower method (because more labour intensive) than others.

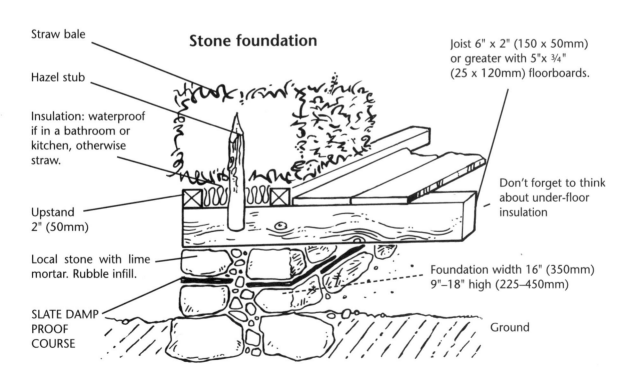

Stone foundation

Straw bale

Hazel stub

Insulation: waterproof if in a bathroom or kitchen, otherwise straw.

Upstand 2" (50mm)

Local stone with lime mortar. Rubble infill.

SLATE DAMP PROOF COURSE

Joist 6" x 2" (150 x 50mm) or greater with 5"x ¾" (25 x 120mm) floorboards.

Don't forget to think about under-floor insulation

Foundation width 16" (350mm) 9"–18" high (225–450mm)

Ground

Simple Blockwork Foundation

This is often a good choice for a cheap and cheerful building

Advantages:

- It's quick and easy to build even with no previous experience

- It is relatively inexpensive and recycled blocks can be used

Disadvantages:

- It's ugly!

- It won't biodegrade into anything useful at the end of its life.

- There is constant potential for damp problems between the concrete block and whatever is above it, as concrete is a 'wet' material and draws moisture into itself.

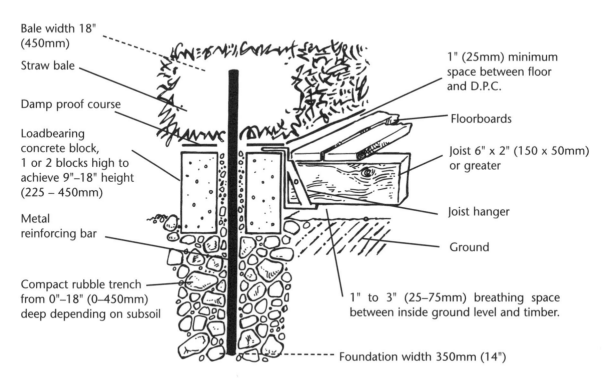

Bale width 18" (450mm)

Straw bale

Damp proof course

Loadbearing concrete block, 1 or 2 blocks high to achieve 9"–18" height (225 – 450mm)

Metal reinforcing bar

Compact rubble trench from 0"–18" (0–450mm) deep depending on subsoil

1" (25mm) minimum space between floor and D.P.C.

Floorboards

Joist 6" x 2" (150 x 50mm) or greater

Joist hanger

Ground

1" to 3" (25–75mm) breathing space between inside ground level and timber.

Foundation width 350mm (14")

Blockwork foundation

Pier Foundations

An excellent example of low-impact foundations and especially useful when building on a sloping site because:

Advantages:

- It can easily cope with different heights of ground by simply increasing the height of the post or pier.

- It is low cost. Using a series of posts or piers is far less expensive than building strip foundations of any description or a 'raft'. (Strip foundations are a trench and/or plinth that are built to the size and shape of the building, below the walls. A concrete ramp is a solid slab of concrete—foundations and floor in one.)

- It has a low-impact on the environment. A series of holes is less intrusive than a trench or ramp beneath the whole house.

- It can create a useful basement space beneath the house, keep the floor above well ventilated, and can also provide an accessible place for plumbing etc.

- It is relatively easy to construct: no special knowledge is required. And depending on materials used, it could be recyclable at the end of its life.

Disadvantage:

- It may limit the design choice in some cases.

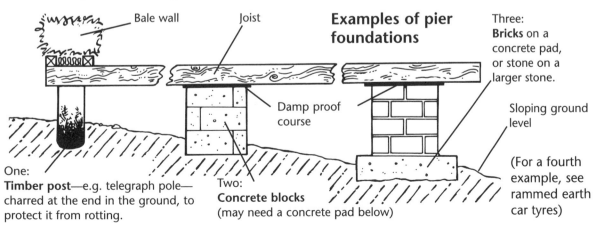

Bale wall Joist

Examples of pier foundations

Three:
Bricks on a concrete pad, or stone on a larger stone.

Damp proof course

Sloping ground level

One:
Timber post—e.g. telegraph pole—charred at the end in the ground, to protect it from rotting.

Two:
Concrete blocks (may need a concrete pad below)

(For a fourth example, see rammed earth car tyres)

Poured Concrete with Slab (or Raft)

This is a method which became popular in the 20th century and is still used because:

Advantages:

- It is a standardised method that most builders are familiar with.

- If done according to the guidelines in the Approved Documents, no extra thinking or discussion needs to be done.

- It is quick and straightforward if machinery is used. Once the preparation is done it can usually be laid in a day, particularly useful on a large building site.

- It means you quickly have a floor surface to work from.
- You can incorporate all fixing points for doors and allow for services such as plumbing and electricity. Of course, this becomes a huge *dis*advantage if you forget . . .

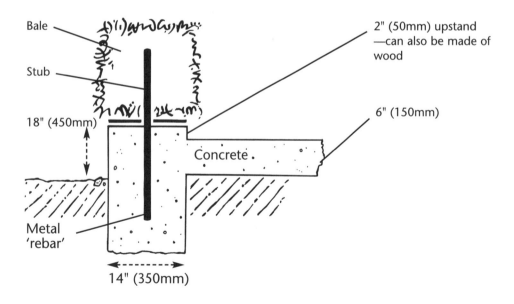

Bale

Stub

18" (450mm)

Metal 'rebar'

14" (350mm)

2" (50mm) upstand —can also be made of wood

6" (150mm)

Concrete

Poured concrete foundation

Disadvantages:

- It creates an enduring damp problem at the interface between the straw and the concrete. Even the use of a dpc at this point doesn't entirely eradicate this. It protects the straw from the constantly wet concrete, but then the dpc itself creates a waterproof surface on which any moisture in the walls will collect! A self-draining foundation is a much better design. Raising the bales up from the dpc on a timber plate can be a partial solution.

- It is costly on the environment because cement takes a lot of energy to produce and to transport, and then at the end of it's usefulness leaves material that does not biodegrade.

- It costs more than you think. There is a popular misconception that laying concrete is cheap—but there is a lot of preparatory work involved to do it correctly. This is increased with a straw building because of the need to raise the foundations above ground level, and to secure metal stubs into them in spacings that correspond to the bales (two per bale, one third of the way along the length of the bale).

- It may be over-designed for the purpose for which it is required

- It's hard, heavy work if you're doing it yourself!

- It's ugly!

Rammed Earth Car Tyre Foundation

This is an excellent choice especially if you have access to a team of volunteers.

Advantages:

- It's very easy to construct. No previous experience is needed.

- It costs almost nothing. Car tyres can be collected for free at most garages.

- It uses materials that are otherwise difficult to dispose of environmentally.

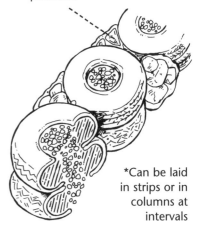

Gaps filled with rocks then plastered

*Can be laid in strips or in columns at intervals

How to place tyres for foundations

- There is no need to use a dpc since the tyres themselves provide this as they are waterproof.

- **It's fun! and very sociable!**

Disadvantages:

- It's labour intensive. This can mean it is costly if you have to pay for labour.

- Ideally tyres should be the same size. It can be hard to sort them out when the garage owner just wants to get rid of everything to you!

- They won't biodegrade into anything useful at the end of their life.

- They are ugly. You will have to plaster them outside so they look better.

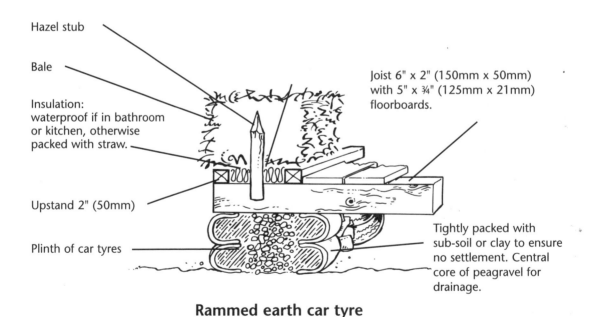

Hazel stub

Bale

Insulation:
waterproof if in bathroom
or kitchen, otherwise
packed with straw.

Upstand 2" (50mm)

Plinth of car tyres

Joist 6" x 2" (150mm x 50mm)
with 5" x ¾" (125mm x 21mm)
floorboards.

Tightly packed with
sub-soil or clay to ensure
no settlement. Central
core of peagravel for
drainage.

**Rammed earth car tyre
foundation**

Foundation Width

The foundation does not have to be as wide as the bales of straw. Straw bales are 450mm (18") wide, but because the edges slope off, the outside 50mm (2") either side do not carry load. This means that the foundation does not need to be more than 350mm wide. Also, once in place, the straw is trimmed off to give a firmer, more even surface for plastering, which reduces the width of the bales. Design will depend primarily upon specific plastering details. **It would not be good practice however, to build foundations that are wider than the straw, as this would encourage water to sit on top of them, and therefore increase the moisture content of the bottom bale.**

Insulation

It is essential to insulate the foundations and floor in some way, so that you don't lose the benefit of super-insulated walls by losing heat through the floor and plinth walls. In areas where there is no risk of plumbing failure and therefore flooding, the space between floor joists can be tightly packed with straw; otherwise more conventional methods such as rockwool or expanded foam types can be used.

Foundations checklist

The above examples of foundation types have all been used successfully with straw bale buildings in the UK and Ireland. It is also possible to use these ideas in combination. What is important is to follow the basic principles:

- **Raising** the bales off the ground (by a minimum 9" [225mm]— preferably 18" [450mm])

- **Securing** the bales to the foundations (preferably with hazel, or alternatively with metal rebar stubs)

- **Raising** the bales at least 1" (25mm) higher than the floor level in any room with plumbing, e.g. kitchen and bathroom.

- **Protecting** the bales from moisture from above and from below.

- **Remembering** to incorporate provisions for good insulation.

Avoid using metal in the walls if at all possible, since it is a cold material and may encourage warm, moisture-laden air from the inside of the house to condense on it.

Chapter 7

Wall Raising

How to do it

For loadbearing

For larger square or rectangular loadbearing buildings, it can be helpful to use temporary corner braces, to provide a guide to keep corners vertical.

Structural doorframes are fixed securely to the foundations or baseplate before the straw is laid. Window boxes are built into the walls as they go up and pinned through the base and sides with hazel.

For framed construction

Depending on the type of frame construction, frames can be built off-site and then assembled once the foundation is finished. All framing including temporary bracing and propping is done before the straw is placed. The roof is also constructed, with felt or tarpaulin and battens to provide waterproof shelter, leaving the final roof covering until the straw is in position, unless the roof covering is very lightweight, such as shingle, in which case no felt is needed.

Lay the base plate onto the foundation if one is being used, and also the floor joists. Fix hazel stubs into the base-plate unless they are already part of the foundation.

Prepare the bales for use (if necessary) by tearing out the centres on each end until a flat surface is created. This ensures that when the bales go together in the wall, there will not be any gaps or large air pockets to reduce insulation. Any small gaps must be stuffed (and not over-stuffed) with straw after every course of bales.

The first course of bales must be placed slowly and carefully as these provide the template from which the walls will emerge. It is important to make sure that the overhang of the bale from the

foundation is correct both sides of the plinth wall; usually the bale is placed centrally. Follow the bale plan accurately.

Bales go together like giant bricks, a second course bale straddling equally the joint between two lower bales. Work from fixed points into the centre of each wall; place the corner bales first, and those beside any framing posts. Bales may need to be handpicked to ensure a snug, not over-tight fit.

Remember to stay calm, work together, and be aware of what other teams are doing on their sections of wall.

Curving Bales

Making bales curve to the shape of a semi-circular design is a highly technical and difficult part of the job. Care must be taken not to laugh too much. The procedure is to turn a bale on its side, lift one end up on to a log, and jump on it! The middle straws in the bale can be moved fairly easily in relation to the strings. Make sure not to curve the bale so much that the string slips off. That's all!

Customising

It will always be necessary to 'customise' bales: to make half bales and bales to fit a specific gap. This can be done easily by using baling needles and restringing both halves of a bale prior to cutting the original strings. A baling needle is a simple tool rather like a giant darning needle, but with 2 holes in the end, to take the 2 strings of the bale. A handle is bent on the other end for ease of use. Attempts to do this with a baling machine beforehand have not been successful. It is difficult in practice for a machine to make uniform sized bales, and the shorter they are, the harder it is. With practice, it can take two people five minutes to customise a bale—a very fast process!

Always customise bales to be slightly smaller than you expect. This allows for the tendency, whilst suffering from bale frenzy, to want to force your new bale into the gap, because you've just spent time making it. And because of the flexibility of straw, this is possible. However, this will almost always result in a distortion of the wall

A baling needle , 20"–22" (500–550mm) long

somewhere else, usually at the nearest corner, or in the buckling of a framing post for a window. Do not give in to the temptation to go for speed rather than a snug fit. Watch out for your work partners and encourage them to adopt a calm and measured approach too!

Bale frenzy: a sort of over-excitement caused by inspirational moments with straw—becomes apparent in any group as soon as the speed with which walls go up is grasped!

Pinning

For Loadbearing

At every radical change of direction, such as at corners, the bales need to be pinned together with hazel hoops or staples. These can be made from 900mm (3') lengths of hazel, 25–32mm (1–1¼") in diameter by splitting the fibres apart with a heavy hammer—without breaking them—and then twisting and bending.

Internal pinning

Once the walls are 4 bales high, they need to be pinned with lengths of hazel. The pins give the wall integrity, so that each bale acts together with the others instead of independently. They are as long as the height of 4 bales, less 50mm (2") which is 1.38m (4' 6"), and they should be 38–50mm (1½"–2") in diameter, straight, sharpened at the narrow end and without excessive knobbles. There are 2 pins per bale (dividing the bales into equal thirds) driven down through the centre of the bale to overlap with the hazel stubs that stand up from the foundations. The same length pins are used in the 5th, 6th and 7th courses too, building up a series of overlapping pins throughout the wall system. The walls of a single or ground floor are usually either six or seven bales high, depending on the design of the foundations and the type of floor installed. First floors are generally from three to five bales high, but can be higher.

For Framed Type
External pinning

Here the pins run from base-plate to wallplate in one continuous piece. (Remember to try and trim the straw ready for plastering

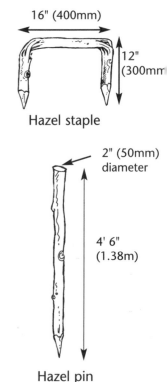

16" (400mm)

12" (300mm)

Hazel staple

2" (50mm) diameter

4' 6" (1.38m)

Hazel pin

before the pins are put in position—it makes the task of trimming quicker and easier). The pins are placed externally to the straw, again two per bale inside and outside the wall, e.g. about 350mm apart and opposite each other. Grooves are cut into the straw with a tool such as the claw on a hammer so that the pins are flush with the straw. Pairs of pins on either side of the wall are tied together through the straw at each course of bales with baling twine, and are fixed to the base and wallplates with screws or nails. The pins are covered with hessian to provide a key for the plaster, either before or after placement. The pins can either be hazel or sawn softwood.

Wallplate or Roofplate

This is a continuous, rigid, perimeter plate that sits on top of the straw bale walls. It is usually made beforehand in sections, for ease of installation, which are fixed securely together once in position. The size of timbers used will depend on the loading it will carry from the roof, the span of the building etc. The foundations or base plates can provide a good template for the wallplate.

Other types of design than the ones illustrated can be used—for instance, a plate that is located at first floor level can also incorporate the floor joists, so as to save on timber.

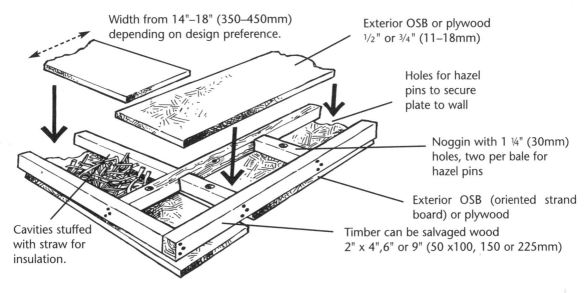

Width from 14"–18" (350–450mm) depending on design preference.

Exterior OSB or plywood ½" or ¾" (11–18mm)

Holes for hazel pins to secure plate to wall

Noggin with 1 ¼" (30mm) holes, two per bale for hazel pins

Exterior OSB (oriented strand board) or plywood

Cavities stuffed with straw for insulation.

Timber can be salvaged wood 2" x 4",6" or 9" (50 x100, 150 or 225mm)

LIGHTWEIGHT FRAME WALL

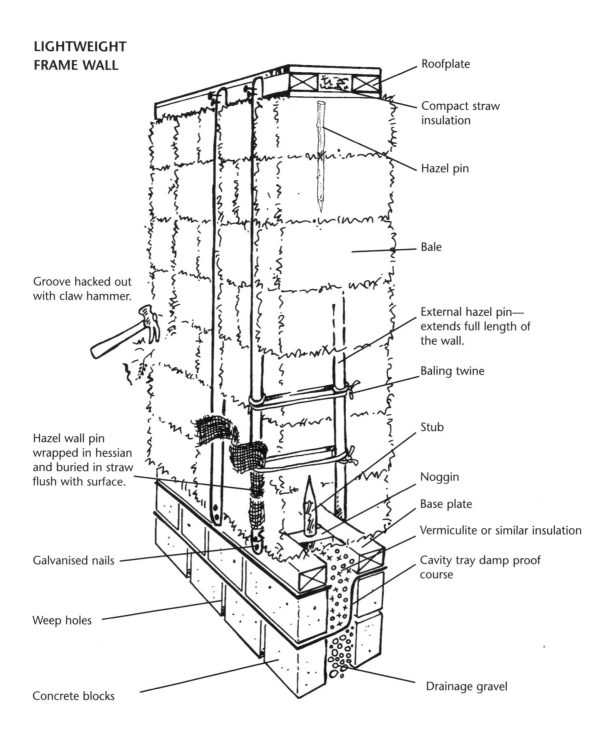

Roofplate

Compact straw insulation

Hazel pin

Groove hacked out with claw hammer.

Bale

External hazel pin— extends full length of the wall.

Baling twine

Stub

Hazel wall pin wrapped in hessian and buried in straw flush with surface.

Noggin

Base plate

Vermiculite or similar insulation

Galvanised nails

Cavity tray damp proof course

Weep holes

Concrete blocks

Drainage gravel

Section through a
LIGHTWEIGHT FRAME WALL
(with loadbearing second storey)

Section through
LOADBEARING WALLS
(1st and 2nd storeys)

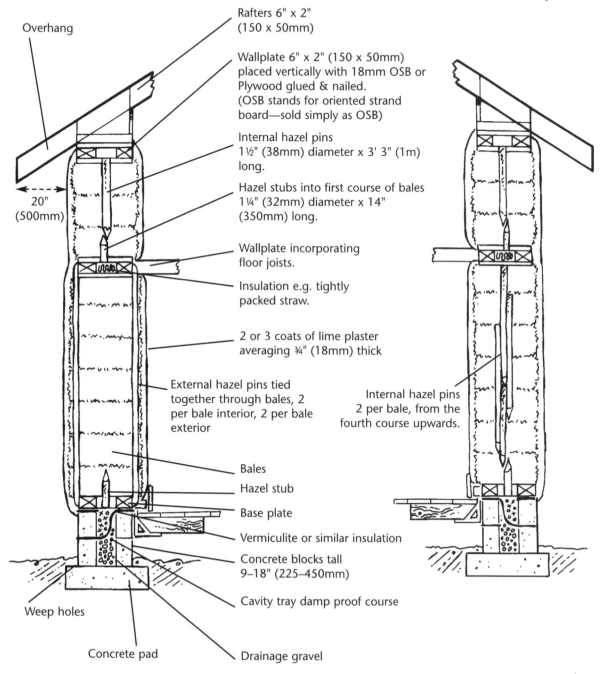

Overhang

20"
(500mm)

Rafters 6" x 2"
(150 x 50mm)

Wallplate 6" x 2" (150 x 50mm)
placed vertically with 18mm OSB or
Plywood glued & nailed.
(OSB stands for oriented strand
board—sold simply as OSB)

Internal hazel pins
1½" (38mm) diameter x 3' 3" (1m)
long.

Hazel stubs into first course of bales
1¼" (32mm) diameter x 14"
(350mm) long.

Wallplate incorporating
floor joists.

Insulation e.g. tightly
packed straw.

2 or 3 coats of lime plaster
averaging ¾" (18mm) thick

External hazel pins tied
together through bales, 2
per bale interior, 2 per bale
exterior

Internal hazel pins
2 per bale, from the
fourth course upwards.

Bales

Hazel stub

Base plate

Vermiculite or similar insulation

Concrete blocks tall
9–18" (225–450mm)

Cavity tray damp proof course

Weep holes

Concrete pad

Drainage gravel

57

Reasons for Using Wallplates

- To evenly distribute the load of the roof or floor across the width of the wall, and around the perimeter of the building.

- To provide a rigid perimeter plate that affords compression of the straw walls at an even rate around the whole building.

- To provide a fixing point for strapping or anchors to the foundation in order to hold the roof structure down against wind uplift, and the fixing point for the rafters themselves.

Once the wallplate is in position, any distortion in shape the walls have suffered due to their flexibility, or bale frenzy, can be adjusted. The weight of the plate immediately gives the walls greater stability. Beginning at the place of best fit, the walls should be persuaded back into correct alignment, and the plate pinned down with hazel 25–38mm (1½"–2") in diameter, 600mm (2') long or longer, again two pins per bale.

Persuader

Settlement and Compression

Ideally we would choose the most dense bales to build with, in order to reduce the amount of settlement that occurs due to the loading of other bales (and floors) and roof. The best building bales will compress by between 12 and 50mm (½"–2") in a 7 bale high wall. Windows and doors in load bearing systems should therefore have a 75mm (3") settlement gap left above them. During settlement, this gap is maintained by folding wedges of timber that gradually reduce the gap as the building compresses. These wedges would be used in all places where settlement needed to occur.

It is also possible to precompress the walls—especially so in the compressed frame method—by using ratchet straps at 1m (3' 3") intervals along the wall, fastened to or through the foundation, to give even pressure on the walls, using the wallplate to spread the load across the width and length of the wall.

At this stage, there is a dramatic change in the stability of the walls, and instead of being flexible stacks giving the impression of a ship at sea, the walls become remarkably solid and reassuring to work on.

Tie-downs

The tie-down is a means whereby wallplates and roof structure can be fastened down securely to the foundations, to prevent uplift when subjected to strong winds. They are usually either metal or plastic, with an adjustable fastener so they can be tightened as the walls compress. They can be fixed once the wallplate is pinned down, or you can wait until the roof has been constructed. It is often easier to trim the walls before the tie-downs have been attached.

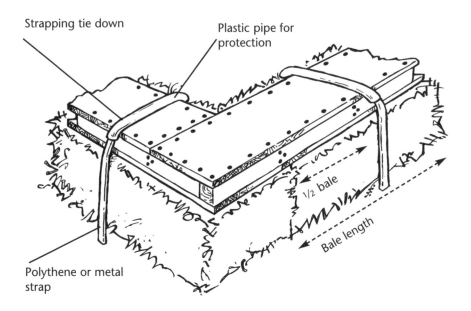

Strapping tie down

Plastic pipe for protection

½ bale

Bale length

Polythene or metal strap

Roof

The design of roof for a straw bale house is not unusual nor particularly different to that for any other building. The main consideration for loadbearing and compressed frame design is that the loading is spread as evenly as possible around the perimeter walls and across their width. This must be remembered whilst the roof is being covered in. Truss rafters should be spread across the walls, not stored at one end of the building before fixing. As the roof is loaded up with slate, tile etc., these should also be distributed evenly and not loaded in one spot, nor should half the roof be slated before the other half.

Whilst any type of roof covering can be used, as long as basic design principles are followed, there are some choices which particularly complement a straw bale house.

Using cedar or oak shingles has to be one of the best choices for environmental and aesthetic reasons. They are a naturally sustainable material and do *not* require a waterproof membrane such as roofing felt underneath them as they are breathable and rely on good ventilation for their extremely long life. They are thus a totally compatible roof covering for equally breathable straw bale walls.

Wheat straw (or reed) thatch is also a great choice as a roof covering. Again, it is a totally natural, renewable and beautiful material. However, both of the above are labour-intensive.

Natural living roofs also complement straw houses, particularly the more modern versions which use only 1" (25mm) of a gravelly soil, with shallow-rooted plants like sedum or strawberries growing in them.

Overhang at the eaves

Straw houses need a good hat to protect them from the weather. A large overhang is a feature of straw bale buildings, especially in this climate. Just as traditional thatched houses have a roof overhang of about 500mm (20"), so too do straw ones. This gives really good protection to the top of the walls against the rain.

Electricity and Plumbing

Again, no real differences in installation. Electricity cables should be encased in plastic conduit sheathing to give extra protection for the (as yet unresearched) theoretical risk due to heat generated by electric cables sited in a super-insulated wall such as straw. They can be buried in the straw and plastered over.

As far as possible, water carrying pipes should be designed to be fixed in internal, non-straw walls, to minimise the risk of water seepage to the straw in the event of a leak. Water-carrying pipes that pass through straw walls should contain no joints, and be encased in larger plastic pipes for the full width of the wall.

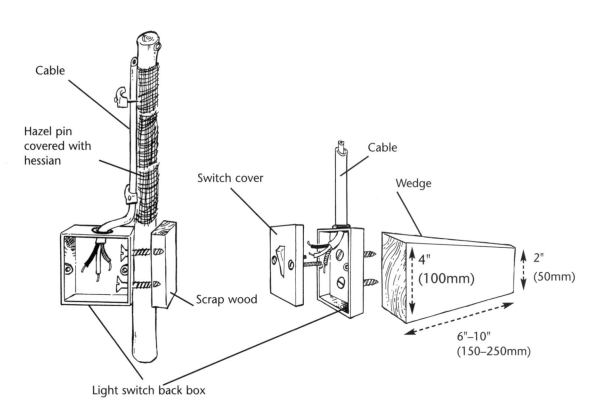

Alternative methods of fixing into straw walls

Internal Fittings

Cupboards, shelves, light switches and sockets, bathroom facilities etc can all be fixed by using timber wedges knocked into the body of a bale, that provide fixings for screws or nails. A sharpened pin of hazel can also be used. These fixing points need to be placed before internal plastering, but can be added at a later stage if necessary. (They can be located after plastering by the simple placing of a nail beforehand.) In framed construction, the framing posts can be used as well.

Chapter 8

Windows and Doors

For Loadbearing

All window and door openings in loadbearing houses must have some way of supporting the weight of the bales, floors and roof over the top of them. Due to the flexibility of straw, the use of concrete or steel lintels is inappropriate and in fact would create problems—the loads need to be spread over as wide a surface area as possible.

A simple way of dealing with openings is to make a structural box frame into which the actual window frame or door frame is fixed.

The design of these box frames must take into account the fact that the straw walls will settle under the weight of the floors and roof above. It is impossible to know how much settlement will occur as it depends on the density of the bales and the amount of loading applied to them. In practice, 75mm (3") is usually sufficient, and the frames are built to be 75mm less than the height of a whole number of bales.

Except in unusual circumstances, structural frames should be multiples of bale dimensions. So external dimensions of the frame could be anything from half a bale to 3 bales in width and any number of bale heights minus 75mm (3") to allow for compression or settlement. Frames are pinned into the bales with hazel, through the base during construction, and through the side once settlement has finished. Door frames would not have the base box as shown below for a window. Instead, the sides of the frame would stand directly on the foundation and be fixed in position with bolts.

The actual sizes of timber used, particularly for the top of the box, will depend on what weight it has to carry. This will be affected by the design of the wallplate above it, which may be able to partially act as a lintel for the window/door. See page 66 for more details of lintel design.

Weatherproofing details around windows and doors are very important. Usually, structural bottom frames should be placed on top of a damp-proof membrane that protects the horizontal surface of the bale below it from damp. The join between the box frame and straw, and between the window frame and box, must be more than adequately covered with plaster and/or timber to prevent draughts and the possibility of rain penetration. Rather than using expanded metal lathing (a 20th-century solution) we

Structural box frame for windows
(box frame for doors is the same minus the base box)

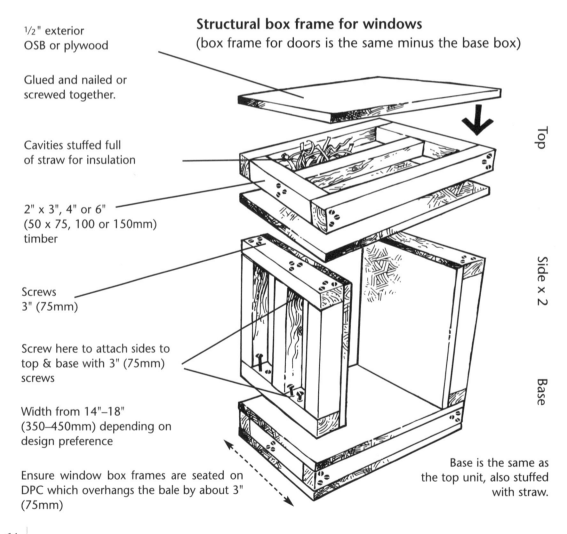

1/2" exterior OSB or plywood

Glued and nailed or screwed together.

Cavities stuffed full of straw for insulation

2" x 3", 4" or 6" (50 x 75, 100 or 150mm) timber

Screws 3" (75mm)

Screw here to attach sides to top & base with 3" (75mm) screws

Width from 14"–18" (350–450mm) depending on design preference

Ensure window box frames are seated on DPC which overhangs the bale by about 3" (75mm)

Top

Side x 2

Base

Base is the same as the top unit, also stuffed with straw.

would use hessian sacking, tacked securely to any timber, which can easily be plastered over to make a weather-tight join between straw and timber.

Framework Methods

In framework methods, windows and doors have upright posts either side of them that run from the base plate to the wallplate above. These posts can be of various designs. A post and beam style would use solid timber and a lightweight frame would use posts slotted at the top to take the wallplate. The framing sill is fixed only after the straw below it has been placed and compressed manually, with a damp-proof membrane below the frame.

In this method, the windows and doors do not need to be multiples of bale lengths, but the design should ensure that the gap between one fixed post and the next does relate to full or half bale lengths.

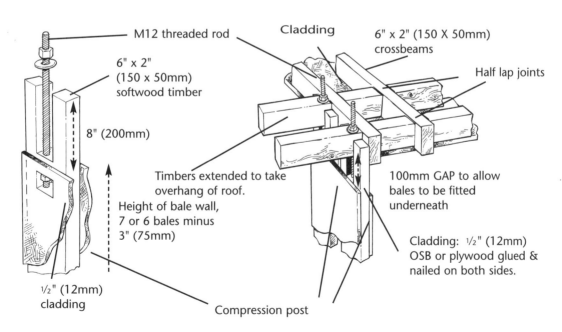

Slotted posts and wallplate design for lightweight frame method

If there is a bale between the top of the window and the wallplate, framing must be designed to carry the full width of the bale, and in the lightweight frame method, allowance should be made for settlement of the wallplate into the slotted posts.

Other Options

1. Use an angle-iron lintel or wooden ladder
This is a ladder welded together from angle iron, using cross pieces to form a cradle into which the bale can sit. It must extend a minimum of half a bale width either side of the opening to spread the load.

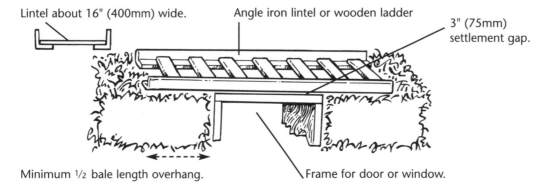

Lintel about 16" (400mm) wide.

Angle iron lintel or wooden ladder

3" (75mm) settlement gap.

Minimum ½ bale length overhang.

Frame for door or window.

CAUTION
In general, metal would not be used in the walls as it may encourage condensation of moisture vapour, as it moves from the interior of the house to the outside. If it is used, it should be covered in an insulating material such as pipe insulation to protect the straw.

2. Attach the window and door box frames directly to the wallplate above it. This can be a good option if the design calls for tall windows, but a settlement gap must be left **below** window frames in this case. It has the advantage of reducing the amount of timber required for the frame. However, it is probably only a first choice option on the ground floor of a 2-storey house, as otherwise the overhang of the roof would obscure much of the light.

Chapter 9

Plastering and Rendering

Straw is a breathable material. It allows the imperceptible passage of moisture vapour and air through it. If it is sealed by a waterproofing material, it will eventually start to rot.

Straw needs good ventilation around it to stay healthy. In practical terms, this means that anything used to weatherproof or decorate the straw must not impede this breathable nature. The ideal finishes for straw are traditional lime based plasters or natural clay plasters, since these are also breathable materials, painted with lime-wash or breathable paints. Clay and lime plasters work well together.

Background to the Use of Lime

Lime has been used as a binding material (mortar) between stone and brick and as a surface protector of buildings (called render when used outside, and plaster when used inside) for thousands of years. All European countries used lime for building, hundreds of years before cement was invented. In the UK and Ireland lime burning was a cottage industry, with local lime pits wherever they were required, and most communities had a working knowledge of its uses and how to produce it. There is no doubt that lime plasters and renders are durable and efficient, well able to do the job of protecting our buildings from the weather.

So, we don't need to argue the case for the ability of lime to withstand the tests of time, weather, and function. However, lime requires thought and understanding of the processes involved in

Imagine putting a bale of straw into a plastic bag and sealing it up. It will start to get hot and sweaty as anaerobic bacteria flourish.

the slow carbonation back to its original limestone, in order to use it successfully. Whilst it is true that a carefully applied lime render or plaster can last for hundreds of years, there have been instances of spectacular failure, and the reasons for these need to be understood if we are not to repeat those mistakes. In essence, the preparation and practice of limework is simple, but variables in the materials themselves, the sand, the lime, and particularly in the weather during application and drying time, are crucial to the overall durability of the material.

Traditionally, knowledge about lime was passed down from one generation to the next, and people were used to using it continuously, and so built up a wealth of experience based on a sound knowledge of the material. Today, there are very few skilled craftsmen (we haven't found any women yet) who worked in those times, and we are having to learn as best we can from what we have left, and remembered histories.

To some extent, what that can lead us into is an over technical approach to what was essentially a practical and rather ad hoc building practice. We are trying to specify exact lime/sand mixes when most likely what happened on site was fairly rough and ready, except for the most prestigious jobs. And mostly, it worked! As tens of thousands of houses in the UK and Ireland, hundreds of years old, can testify. So what follows is an attempt to explain what happens in the lime burning, slaking, mixing process, and what is important to know, so that you can take care of your own limework satisfactorily.

Limestone and Lime-burning

The raw material for all lime mortars and renders is naturally occurring limestone, shells or coral, which is called calcium carbonate ($CaCO_3$). It is made into lime putty by a relatively simple process. Traditionally, the limestone is placed in a specially built kiln (sometimes a pit or a heap) and layered with fuel such as coal or brush and burnt for about 12 hours. It needs to reach a temperature of 900–1200 °C; 900 °C for carbon dioxide (CO_2) to be driven off, and 1200 °C for the heat to penetrate through to the centre of the stone.

As it heats up, steam is driven off first (water, H20), which is always present in the limestone, and the following chemical change takes place:

heat + $CaCO_3$ = CaO + CO_2
(heat + calcium carbonate = calcium oxide + carbon dioxide)

At the end of the burning process, whitish lumps of calcium oxide are left with bits of burnt and unburnt fuel. Over-burnt limestone appears as black, glassy pieces, and these should be removed and discarded. The chemical reaction that takes place is usually more complicated than this, due to other carbonates and silicates being present in the limestone, but it's important to understand the basic changes that are taking place at this stage. Calcium oxide is very reactive and can be dangerous; it is called "lump-lime" or "quick-lime" and may be left as lumps or ground down into powder. It MUST be kept dry as it reacts very quickly with water, even the water in the air or the moisture in your skin, to form calcium hydroxide, which is the first step to reversing the process back to calcium carbonate again. Just as making quicklime needed heat, the reverse process PRODUCES heat.

CaO + H_2O = Ca $(OH)_2$ + heat
calcium oxide + water = calcium hydroxide

So quicklime added to water gives us . . . lime putty!

Caution: Never add water to quicklime. Always do it the other way round and add quicklime to water, or else it could explode!

Protective clothing, goggles and mask should be worn when working with quicklime.

How to Make Lime Render and Plaster

There are two main ways to do this

1. Lime putty mix

Recipe: 1 part lime putty to 3 parts sand

The sand MUST be well graded and sharp, that is, contain particle sizes ranging from very small (dust) to quite large (5mm or 3/8"), and these should be angular not rounded. When compressed together, the aim is to use as much lime putty as necessary to fill the spaces between the grains (the VOID spaces) but no more. The mix is almost always 3 parts sand to 1 part lime putty (3:1) because the void spaces take up about 33% or 1/3 of the volume of most sands.

The only real difference between a plaster (for inside work) and a render (for outside work) is the fineness or coarseness of the sand used. Render may contain aggregate particles up to 10mm in size in some areas that experience lots of wind-driven rain; inside a smoother finish is usually preferred, using a sand of smaller grain size.

The longer a lime putty has matured, the more solid it becomes, and the better render it makes. It may seem hard to work at first, but by pounding and beating it with wooden mallets or posts it soon becomes more plastic and can be worked into the sand. It can be VERY labour intensive, and this beating part should not be missed out. Because it's so hard to work, it can be easier to mix the sand with fresh lime putty, and then leave THIS mix to mature for 3 months, traditionally under a thick layer of sand, and then straw!

Well graded sharp sand, particle size dust to 5mm.

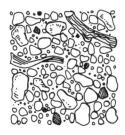

Poorly graded and round sand with organic debris.

2. Hot lime mix

Recipe: 1 part quicklime powder to 3 parts sand

This is probably the most common method used historically for mortars. In this method, the quicklime is added to *damp* sand in a bath and mixed with a shovel. Very soon, the mix starts steaming

For more information on lime, see Reference Section, page 95

and becomes warm, as the reactive calcium oxide hydrates with the water in the sand. At this point, the mix can be riddled (passed through a sieve, usually ¼" for renders), as it's easier to do it when the quicklime has dried out the sand. This process is dangerous because the powdered quicklime blows in the air and can get into eyes and lungs, reacting with the moisture there, plus the mix gets hot very quickly and may be difficult to control. It must be raked and mixed continuously, and depending on the dampness of the sand, may not need any extra water adding. Again, it should ideally be left to mature for at least 3 months.

How to use Lime Render and Plaster

The internal and external faces of the straw walls should be given a very short haircut—trimmed down to a neat finish. All the long, hairy, unkempt bits of straw should be removed.

The reasons for this are:

- To minimise flame spread over the surface of the bales, in the event of a fire before plastering.

- To reduce the amount of plaster required by reducing the surface area.

- To even out any large undulations in the surface of the wall.

Plasters and renders can be bought ready-mixed from one of a growing number of suppliers; they can be mixed on site from lime putty and local sand, or sometimes from quicklime and sand, depending on local availability. Lime putty can be ordered from a supplier, ready-made in airtight sacks. Well-graded sharp sand can be bought from a builders merchant, but keep it clean. If you are making your own plaster, it can be stored in a trench or pit for a minimum of three months, covered with sacking and straw. The lime mix should be applied directly to the trimmed straw.

There is no need to wrap the straw in stucco or chicken wire first. Both lime and clay stick extremely well to the straw, particularly if applied by hand or sprayer.

There is no need to wrap the straw in stucco or chicken wire first, as many cement rendered buildings in the USA have been. It is totally unnecessary and a waste of time! Both lime and clay stick extremely well to the straw, particularly if applied by hand or sprayer.

The lime mortar or render should be beaten and worked to a stiff consistency, so sticky that it can be held upside down on a trowel. There should be no need to add water to it, this would increase the risk of shrinkage cracks. It will generally become more plastic with lots of beating! Traditionally, it was a completely separate trade, to be a lime render beater. These days, render can be knocked up in a paddle mill (used by potters) to save all that work by hand. Generally, a cement mixer WON'T do the job as the mix stays in a lump and knocks the machine over, then the tendency is to add water to soften it, and the resultant mix will crack due to too much shrinkage! If the mortar has been matured by the supplier and then delivered, the vibration caused by the delivery lorry can be enough to soften up the mix. In this case, it may be possible to use a cement mixer to complete the process!

The first (or scratch) coat onto straw is usually lime-rich to make it stickier, often a 2:1 mix. The next 2 coats of plaster contain cow or goat hairs, or chopped fibres such as straw or coir, well distributed throughout the mix, to give it much greater strength; in the same way that straw is used in mud plasters, to give it tensile strength.

For straw bale walls, it's usually best to apply the first coat of lime by hand (with gloves!) because it's more fun, and the straw tends to flick the stuff back at you if you try to use something like a float. It needs to be well rubbed in, to get a good key (join) between the straw and the lime. It's important to encourage the render to cure (go off) from the inside out, not to let the outside skin carbonate too fast: the way this is done is to keep the whole thing MOIST (not wet). The surface should not be allowed to dry out; it will naturally take 2–7 days before the render feels hard. The first coat should be as thin as possible, leaving stubbly bits of straw sticking out, and will probably be ready for the second coat on the next day, unless there are pockets of thicker mix in places. A rule of thumb, literally, is to put the second coat on when the first is hard enough that you cannot push your thumb into it. Wet the walls down with a mister, not a hosepipe, before putting the second coat on, and work it well in, either with hands again or a wooden float. Keep the render damp by misting it, unless you have ideal drizzling weather! Keep going over the wall with a wooden float, rubbing in the mix and misting it. It's probable that lime renders on straw bale walls carbonate more quickly than on stonework, because the straw itself is breathable, and so the back, as well as the surface of the render has access to the air.

Over the next few days, protect the render from direct sunlight, driving rain, forceful wind and frost. Often hanging sacking from scaffolding, and keeping the sacking moist to create a humid atmosphere close to the lime does this. The render WILL crack, and needs to be reworked several times over the next few days to squeeze and compress the sand particles together, before the surface hardens. The cracks are caused by shrinkage as the excess water in the mix evaporates. The aim is to compress all the render so that there are no air spaces left. The misting is not to add water to the render, but to make sure that carbon dioxide can be carried into the thickness of the layer via the medium of water. It needs to be protected from frost for about 3 months so don't do it too late in the year!

It is not a good idea to use a steel float on a lime render, as this polishes up the surface and closes up the texture, thus preventing humid air from penetrating into the body of the render.

It's probable that lime renders on straw bale walls carbonate more quickly than on stonework, because the straw itself is breathable, and so the back of the render has access to the air, as well as the surface.

It needs to be protected from frost for about 3 months so don't do it too late in the year

Limewash and Decorating

Applying limewash to the building once it's been plastered should be seen as part of the plastering process. If there are any tiny cracks left in the finished plaster, the limewash will seal these up. Over time, lime plasters have a self-healing effect. Any cracks that do appear tend to close up as the lime carbonates, because the calcium carbonate molecule is bigger than the calcium hydroxide one. Limewash is made from lime putty, diluted with water. It is very thin and watery, and should be applied in many thin coats, left to dry overnight each time.

Externally, walls that take a lot of weather, usually the south-west side, should have about 5 coats of limewash to protect them. The rest of the building may only need 3, although the more coats you apply initially, the better the weatherproofing will be. How frequently it needs re-coating will depend on the weather. The sheltered side may only need limewashing again every 5 years, whereas other parts may need to be done more often. Natural pigments can be added to the limewash, giving a large range of beautiful colours, which have a qualitative visual difference when compared with modern vinyl paints. Limewashes are more variegated, less stark and uniform. Lovely!

For more information on lime, see Reference Section, page 95

Natural Clay Plasters and Renders

Although these types of plaster are very common in the rest of Europe, Scandinavia, the USA, the Middle East and Africa, they are not so well known in the UK and Ireland. Knowledge of their use has largely been lost, although we do still have many fine examples of older buildings with a clay mortar binding the bricks or stones together. And of course, our rich heritage of cob buildings, built entirely of clay and sand, stand testament to the durability of clay finishes.

Depending on the geology of your local area, you may find a clay sub-soil that is ideal for plastering, or pockets of clay that can be added to sand to make a good render. Clay types differ, but in general, a plaster or render needs about 20% clay to 80% sand.

Clay is applied to straw in the same way that lime is. The first coat onto the straw is rubbed in by hand, and would be a thin, clay rich mix. All other coats would have lots of finely chopped straw mixed into the plaster to give it tensile strength and stop it cracking. The final coat would use finer sand to give a smoother finish.

Often, clay coats would be applied before the lime plaster, to even out undulations and to save on the amount of lime plaster used. It is often used as a finish coat inside a house, but would not generally be used on the outside as a finish except in very sheltered positions. Inside, it may be limewashed or coated with linseed oil etc., according to the effect required. Outside, it would probably require several coats of limewash to protect it from the weather.

There is a clay plaster that is commercially available, imported from Germany. It comes in powder form in sacks, and you just have to add water. It works brilliantly well, and there's no waste as anything that is dropped or dries out can just be remixed with water. Different grades are available for the backing coat and finish coats. At the moment, it is expensive to buy, approximately 4 times as costly as the equivalent amount of sand and cement. **It is an open market: our own brick companies could manufacture home-produced clay plasters at reasonable prices!**

In the meantime you can make your own, but mixing clay and sand from raw materials can be very laborious and time-consuming. It works best if the clay is either completely dry so that it can be powdered, or completely wet so it is a thick slurry. In either form, it can then be mixed with sand using a shovel, in the same way as cement.

Other ways of mixing are:

- To trample the whole mix by foot in wellies or bare feet. This is a lot of fun if done in a group, but takes a long time and can be tiring.

- To use a paddle mill. Potters may have one of these. It's a round pan with heavy wheels inside that turn and squash the clay mix at the same time as the pan is turning and the clay is scraped up off the base. It's the best way of mixing but it can be hard to find one and expensive to hire.

- To use a tractor. This method works very well too, especially for large quantities, but it can take a long time to get all the small lumps out of the mix.

In general, it would not be sensible to mix your own clay plasters except for small buildings, where you have a lot of help, or when you don't have to pay labour costs.

One of the great advantages of taking time to use clay plasters is that it gives a great opportunity for creative expression. The clay can be sculpted and moulded into all sorts of frescoes and reliefs. In fact, it is almost impossible to stop people from being creative with it, it is so tactile and such a lot of fun to apply.

Cement Plasters and Renders

There are hundreds of straw bale buildings in the USA and Canada that have been cement rendered. Most of these are doing fine and are not showing any sign of deterioration. Some of them, however, have become very damp as a direct consequence of using a cement render.

Cement and lime are materials that behave very differently to each other, and are used for different reasons. Whereas lime is a breathable material that holds water within itself whilst it is raining, and releases it once the rain stops, cement is waterproof and works by preventing water from penetrating through it to the surface beneath. Also, lime is quite flexible, whereas cement is rigid. This means that as long as there are no cracks in the cement, it will stop water from reaching the straw. However, due to its rigidity it is almost impossible for it not to crack after a short period of time, especially when it is applied to a flexible backing material like straw. This means that when it rains, the rain passes through tiny cracks and filters down the inside face of the cement, and collects at the bottom of the wall, where it cannot get out. A build-up of trapped moisture at the base of the wall causes the rot to set in.

The other consideration with cement plasters and renders is that because they are *not* breathable they have the same effect as wrapping the straw in plastic. This lack of breathability creates an unhealthy, unventilated atmosphere around the straw which could lead to damp problems in the long term.

In practice, there may be many instances where you can get away with using cement, or where the life of the building is such that a bit of rot developing at the base of the wall does not matter.

There is no doubt however, that in terms of best practice, lime renders are superior to cement.

Chapter 10

Planning Permission

Planning policy is a political subject that is determined broadly at national level, and in specifics at local level. Whilst there are general similarities in all areas, there will be differences in policy locally that reflect local circumstances. However, **the fact that a house is built with straw walls is of very little concern to the planners,** although it will be to the Building Regulation department. The planning department, guided by local elected councillors, will have worked out a comprehensive plan for the area that specifies where new housing can be built, which areas are to be kept as green-belt, which is agricultural land etc. Within each area, different types of building will be allowed or not allowed, according to guidelines that have been set by political considerations. It may well be important to know what the planning policy is for your area, and to understand why the Planning Authority has made these decisions. For instance, if you wish to build a 3-bedroomed house in a local farmer's field in England, you are unlikely to get permission to do so, because the field is probably designated as agricultural land and therefore no domestic buildings will be allowed. However, if you wish to build on a site next to other houses, you probably would get permission.

In Ireland, the situation may be different unless you are near a town, because there is more of a tradition of self-build, and it is much more common for people to build on family land. Also, there are fewer restrictions on securing permission to build because the country is far less populated than the UK.

Most planning decisions are subjective and political, and your planning officer can be of invaluable help in informing you of basic policy, and of particular circumstances in which there may be room for negotiation. It is a sensible approach to recognise your planning

It is advisable to have a good relationship with your planning officer.

officer as someone who has useful knowledge that can be shared with you to enhance your project. An application for planning permission has more chance of success if the planning officer supports it. It is always best to find a way to work together, if at all possible.

Procedure in Ireland differs from that in the UK. In Ireland, the local council takes the decision, taking objections into account, after which building must not commence for one month. During that month, objections to the Council's decision can be lodged, on payment of a fee, with an independent Board. The Board then makes its final decision based on the Council's decision and the arguments of the objectors. Beyond this there is a government appeals system to which you can refer if the decision is unfavourable.

In the UK the local authority makes the decision based on its policies and any objections raised. If the decision is unfavourable it can be appealed through a government process. In many cases, planning permission will not be required either because a building is too small, or because it is within your 'curtilage' (garden area).

Areas of Concern for Planners

What does it look like?

All Local Authorities will be concerned primarily with this question. In general, your building must fit in with local surroundings, it usually has to look similar to others in the locality and not be an eyesore. Of course, what we each define as eyesore can vary dramatically! Some think concrete bungalows are beautifully modern, and others hate them. In areas of the Pennines (in northern England), for instance, all houses must be built of local stone. However, some developers have argued successfully to build out of concrete that looks like stone. And there's at least one straw bale building that has planning permission as long as the outside render is stone coloured. This illustrates really well the possibilities for negotiation that exist within any planning policy.

What will it be used for?

The purpose for building is important. Are you going to live in it, open it as a shop, store machinery in it, hold band practices? Homes usually require access for vehicles and a means of dealing with sewage and waste water.

What you do in it has implications for wider services and the impact you'll make on the social and physical environment. Just because you want to live on a greenfield site and make little impact on the environment doesn't mean the planners will let you. They may be concerned, not about you, but about the owners who come after you when you sell. And just because you think you can deal with your own sewage and waste water doesn't mean that the planners will agree. Besides, some areas will be designated for housing and others for remaining unspoilt. It will be difficult in any area (though not impossible) to cross these boundaries.

What do the neighbours think?

This isn't necessarily as big an issue as it may seem. Planners do have to take into account different viewpoints and in some areas anything new or different will cause a stir but there have to be legitimate reasons in order to object to it. Planners may choose not to contend with a powerful local lobby that has no real grounds for objection, or they may think it's politic not to ignore them. However, negative reactions from the neighbours may simply be seen as emotional responses to change, and positive reactions may help you argue your case for innovative design.

Environmental issues & Agenda 21

Every Local Authority has the duty to implement European directives (Agenda 21) relating to issues of sustainability and protection of the environment. The emphasis these directives are given can vary tremendously from one Local Authority to the next, but in general there is now greater awareness of the need to build using materials and practices that are less harmful to the planet. If your house fulfils some of these directives, the planners may react more favourably to it, even if it differs in some significant way from other planning guidelines. For instance, a plastered straw bale house may be allowed in an area where most houses are brick because although it looks different, it provides three times more insulation—thus reducing dependence on fuel, fossil fuels etc—than equivalent houses in the locality.

Chapter 11

Building Regulations

There is no need to be unduly worried about whether your straw bale house will meet all the Building Regulation Requirements. It definitely can, and almost certainly will! It's important to understand that we 'regulate buildings' in order to make sure that they do not pose a threat to anyone or anything in terms of health and safety. The procedure for dealing with the Regulations is different in Ireland and the UK.

In Ireland the plans, which may or may not include all relevant building regulation information, are submitted for approval to the Local Planning Authority, and everything is dealt with by them. The responsibility for ensuring compliance with the Regulations lies with the person who has overall responsibility for the design and its practical implementation, and who oversees the quality of work. It is expected that this will be an architect, but it doesn't have to be. You may have submitted your own plans, for instance, or your building firm may have designated its own person. The Local Authority may come to inspect your building at any time, but may not.

In the UK the Planning department and the Building Regulation department are separate and require different information, although it is usual for all information to be contained on one set of drawings, submitted to both departments. Ultimate responsibility for ensuring compliance lies with the Building Control Inspector, who works for the Local Authority. They will insist on visually making sure that certain aspects of the construction are built according to plan: for instance they will want to see that specified damp proof courses are actually in place before concrete is poured, etc. You are required to pay a fee for this service and it can be very helpful to have the Inspector on site if you want to ask any questions.

The building regulations for both Ireland and the UK are contained in a number of Documents, called Technical Guidance Documents in Ireland, and Approved Documents in the UK. In Ireland they can be obtained from the Government Publications Sale Office in Dublin, and in the UK from an HMSO bookstore. Whilst there are some differences between them (for instance the Irish Regulations make reference to European Standards on materials as well as British Standards), one of the most notable differences is that the Irish ones are printed on recycled paper and the UK ones are not!

Both sets of Documents are labelled from A to M (the UK set also has a Document N), covering the same subjects.

They clearly state:

> "The adoption of an approach other than that outlined in the guidance is not precluded provided that the relevant requirements of the Regulations are complied with."
> **Irish Technical Guidance Document**

> "The detailed provisions contained in the Approved Documents are intended to provide *guidance* for some of the more common building situations." (my italics)

and:

> "There is no obligation to adopt any particular solution contained in an Approved Document if you prefer to meet the relevant requirement in some other way."
> **UK Approved Document**

'Document A', for instance, refers to the structure of a building and will advise you on the minimum thickness your walls should be, and the thickness of concrete you should have in your foundations. This example immediately highlights a major issue around straw bale building and the building regulations; the regulations are written to cover the most common types of 20th century building materials, that is, concrete, brick and timber. If you are choosing to use other types of materials, or to use the same ones in different ways, then you will have to discuss this with your Builder/Architect in Ireland and your Building Inspector in the UK, because there will be no written guidelines. They do not mention straw walls 450mm thick, built on timber post foundations, for instance. But this does not mean you cannot do it. On the whole, we would expect the Builder/Architect and Building Inspector to be sensible, well-informed people who are up to date with current developments in building practice. They have lots of useful knowledge that can be of help to you in designing your building, and can access their colleagues or other advisors if they need to inform themselves further about any subject.

There are many people in the construction industry who are not aware that there are other ways of building, or who are nervous of stepping away from what they know, and what is written in the Documents.

This does not mean it cannot be done, and in the UK your Inspector is often the best person to help you with this, (together with your straw bale advisor of course!). When contemplating building anything new or unusual it is necessary to go back to first principles and look at what the *aims* of the Regulations are. When a wall has a greater thickness, as a straw bale one does, but weighs a lot less than more common materials such as brick, then it is reasonable to assume that the foundations would not need to be as substantial as for a brick wall in order to provide the same level of stability. The Regulations aim to ensure that whatever is built does not pose a health and safety threat in any way.

The Regulations cover all aspects of building, but for our purposes in the use of straw, the only areas that are substantially different to other types of common 20th century styles are the walls and therefore the foundations. So the areas of concern for Building Regulations are:

- Insulation
- Fire
- Structure
- Durability (including degradation due to moisture)

Thermal Insulation

Nowadays, all new buildings must be energy efficient. This covers many aspects of the building, including design to reduce heat loss. The usual way we do this is by using insulation of one sort or another. In brick or block walls, this often takes the form of an expanded polystyrene or foam stuck to the back of the blocks inside the cavity of the wall. With straw bale walls, the insulation (straw) is also the building block.

The amount of insulation of a material is measured by its U-value.

The U-value, or thermal transmittance, of a material is the amount of heat transmitted per unit area of the material per unit temperature difference between inside and outside environments.

It is measured in units of Watts per square metre per degree of temperature difference (usually measured in Kelvin) W/m^2K. Put simply, it's a measure of how much heat a material allows to pass through it.

The lower the U-value, the greater the insulation of the material.

Building Regulations currently require that external walls of domestic dwellings must have a U-value of 0.35 or less.

Straw bales, because their width is 450mm (18"), have a U-value of about 0.13.

The high insulation value of straw is achieved because of the width of the bales.

Compare the U-values of other common wall building materials:

105mm brickwork, 75mm mineral fibre,
100mm light concrete block, 13mm lightweight plaster: **0.33**

100mm heavyweight concrete block, 75mm mineral fibre,
100mm heavyweight concrete block, 13mm lightweight plaster: **0.40**

100mm lightweight concrete block, 75mm mineral fibre,
100mm lightweight concrete block, 13mm lightweight plaster: **0.29**
(CIPSE: Thermal Properties of Building Structures)

Unplastered straw has a U-value of about 0.13; the addition of plaster would bring this even lower, so there is no doubt that straw bale walls exceed the requirements of Building Regulations for thermal insulation. Further information on the thermal tests conducted so far on straw bales is available in the Reference Section, page 103.

Measured bale conductivity results

Research by:	Joe McCabe	Joe McCabe	Sandia Labs	ORNL	CEC/ATI	ORNL	Ship Harbour
Year	1993	1993	1994	1996	1997	1998	1995
Procedure	hot plate, single bale	hot plate, single bale	Thermal probe, single bale	hot box full wall	hot box full wall	hot box full wall	heatflow monitoring
Disposition of bale	on flat	on edge		on flat	on flat	on flat	on flat
Straw type	wheat		wheat	rice		wheat	wheat, oats, barley
Bale width	0.58m: 3-string		0.46m: 2-string	0.46m: 2-string	0.58m: 3-string	0.48m: 2-string	0.46m: 2-string
Moisture content	8.4%		not given	not given	11%	13%	variable: 8% – 15%
Density / kg m^{-3}	133	133	83	not given	107		128
Conductivity / Wm^{-1}K^{-1}, Known as the 'K' value	0.061	0.054	0.054	0.153	0.128	0.100	0.100
Predicted U-value, in Wm^{-1}K^{-1}, for 2-string bale, stuccoed on both sides	0.13	0.15	0.12	0.31	0.26	0.21	0.21

Information compiled by Mark Bigland-Pritchard hyphen@dial.pipex.com. Mark thinks that the CEC test and the first ORNL test included design faults that resulted in unreliably high k-values, and so would disregard these.

Sound Insulation

New regulations are under discussion, covering sound insulation of party walls in order to make homes quieter. There are, as yet, no official research findings for quantifying the level of sound insulation provided by straw bales. However, we have overwhelming experiential evidence that straw walls offer far more sound insulation than 20th century wall building techniques. People who live in, use or visit straw bale buildings remark on the quality of atmosphere found inside one. They are cosy, calm and quiet. They offer a feeling of peace. There are at least two sound studios in the USA built of straw because of its acoustic properties, and several more meditation centres (see back issues of *The Last Straw*). Straw bale walls are increasingly being used by airports and motorway systems as sound barriers to reduce traffic noise.

Fire

There is no question that straw bale walls fulfil all the requirements for fire safety as contained in the Documents.

Research in the USA and Canada conclude that:

"The straw bales/mortar structure wall has proven to be exceptionally resistant to fire. The straw bales hold enough air to provide good insulation value but because they are compacted firmly they don't hold enough air to permit combustion."—Report to the Canada Mortgage and Housing Corporation. Bob Platts 1997

"ASTM tests for fire-resistance have been completed... The results of these tests have proven that a straw bale infill wall assembly is a far greater fire resistive assembly than a wood frame wall assembly using the same finishes."—Report to the Construction Industries Division by Manuel A. Fernandez, State Architect and head of Permitting and Plan Approval, CID, State of New Mexico, USA.

Straw bale walls are less of a fire risk than timber frame walls.

It is a popular misconception that straw bale buildings are a fire risk. This misconception seems to come partly from the confusion of straw with hay, and the collective memory of (relatively rare) spontaneous combustion in hay barns (from large haystacks baled too wet and green). Straw is a very different material to hay, and there are no known cases of spontaneous combustion with straw, even when stored in poor conditions.

There is a risk of fire with straw during the storage and construction process. It is loose straw which is the risk, since it readily combusts. If you were to cut the strings on a bale and make a loose pile of the straw, it would burn very easily, as it contains lots of air. Therefore it is essential to clear loose straw from the site daily, store straw bales safely, have a no-smoking policy on site, and protect the site from vandalism. If the wall is to be unplastered for a while, be sure to trim it, getting rid of the 'fluffy bits' which would encourage flame spread.

Once the straw is built up into a single bale wall it tends to behave as though it were solid timber, particularly when it is loadbearing, but also when used as infill. In a fire, it chars on the outside and then the charring itself protects the straw from further burning. It's like trying to burn a telephone directory—if you tear loose pages from it, they will burn easily, but if you try to set fire to the whole book, it's very difficult.

When the wall is plastered both sides, the risk of fire is reduced even further, as the plaster itself provides fire protection.

For the purposes of building regulations, a wall built of *any* material that is covered with half an inch of plaster has a half hour fire protection rating, which is the requirement for domestic buildings. All the fire-testing research done on straw bale walls, concludes that this type of wall-building system is *not* a fire risk. A list of research documents can be found in the Reference Section on page 103.

Structure

The requirements laid down in 'Document A: Structure', are for brick, concrete or timber walls. You will find no guidance here for building straw bale walls. This does not mean it cannot be done!

Research in the USA has shown that structural loadbearing straw bale walls can withstand loads of more than 10,000 lbs/sq.ft, equivalent to 48,826 kg/m². *Research by Ghailene Bou-Ali: Results of a Structural Straw Bale Testing Program 1993.*

There is no doubt that loadbearing straw walls can withstand greater loads than will be imposed on them by floors, roofs and possible snow loading. It is the design of associated timber work, the even spread of loads around the walls, and the quality of building which is crucial here, *not* whether the straw can do it.

With infill walls, in post and beam type structures, the straw does not take weight anyway and there are conventional calculations available for structural strength of other types of framing.

Durability

This is the area of most concern when designing straw bale houses in order to comply with Building Regulations.

Will the straw bale walls retain their structural integrity over time, or will they suffer material degradation caused by moisture, either from condensation, rain or ground water? Whilst this is a consideration for all house builders, in fact all building regulations require is that the walls pose no threat to health and safety. There has been no research so far on the durability of straw bale houses in the weather conditions we experience here. What little research has been done in the USA and Canada shows that there should be no *need* to be concerned that straw bale walls will not withstand the test of time and the rigours of our climate. The key to durability lies in good design, good quality work and maintenance. Past experience is an allowable and viable method of establishing the fitness of a material as it says in the Documents, if:

"The material can be shown by experience, such as its use in a substantially similar way in an existing building, to be capable of enabling the building to satisfy the relevant functional requirements of the Building Regulations."

Irish Technical Guidance Document

The key to durability lies in good design, good quality work and maintenance.

"The material can be shown by experience, such as in a building in use, to be capable of performing the function for which it is intended."

UK Approved Document

There is also a specific reference to the use of short-lived materials in the UK Documents:

"A short-lived material which is readily accessible for inspection, maintenance and replacement may meet the requirements of the Regulations provided that the consequences of failure are not likely to be serious to the health or safety of persons in and around the building."

Not that straw *is* a short-lived material, but this clause should reassure anyone who is still not fully convinced of the capabilities of straw. In any case, a building that is designed well and built well should not experience any long term effects of degradation due to moisture. There are plenty of examples in the USA of straw bale houses enduring for over 50 years with no signs of deterioration. However, it is true to say that our experience of building in Ireland and the UK is only 7 years old, and we do not have evidence such that we can state with certainty that straw bale buildings will survive for long time periods in our climate. We do, though, know that *even if* there is degradation of the straw, it:

• is easily repaired

• degrades slowly and therefore poses no risk to safety.

Finally, a word of caution about Building Codes, Building Regulations and building practice in general.

You need to be careful about what you read in books and on the internet about straw bale building and how it must be done. Most of the information available up to now is based on American Building Codes and methods of building, which are not necessarily appropriate for us in Ireland and the UK. There is a fundamental difference between the USA Codes of practice and our Building Regulations:

In the USA, Codes are prescriptive: that is, they tell you that you MUST do it this way. In Ireland and the UK, Building Regulations are guidelines. They advise you on best practice, but you can do it another way if you can show it's effective.

Prescriptive codes can mean that as new and simpler techniques are developed and the nature of straw is understood more fully, new practices cannot be used officially in the USA until the Building Codes have been altered. This was highlighted by the practice of using stucco wire in cement renders. The Codes in some States insist that stucco wire must be used on straw buildings that are cement rendered. This means that even if you are sure it is not necessary, you still have to do it in the USA. Very quickly it becomes 'fact' that straw bale buildings must be wrapped in stucco wire, when the reason for the 'must' has become lost. So it is healthy to practice common sense, coupled with an enquiring mind.

When faced with a choice about whether or not to try a new or different construction technique, always ask yourself first: *does it work*? But don't stop there. Ask the most important question next: *is it necessary, or does it just over-complicate the building*? The best straw bale homes are straightforward and beautifully simple.

Frequently Asked Questions

What about mice and rats?

There is no greater risk of encouraging mice and rats into your straw bale house than there is for any other type of building. Straw is the empty stem of a baled grain crop and unlike hay, it doesn't contain food to attract furry creatures. Any home where food is left out in the open is a potential lure for vermin. Once your straw bale house is plastered, the walls seem no different to a mouse than other plastered walls. Mice and rats like to live in spaces between things, as they are very sociable animals. In barns, they live in the gaps between bales and in houses they live in cavities and under floors. If you build straw walls and then clad them in timber, with an air gap between, this might attract mice: but it's the gap they like, not particularly the straw. If you build straw walls, plaster them with clay/lime and maintain them, then there are no gaps to invite them in, and no cavities in which they can live.

How long will it last?

No one can completely answer this question because the first straw bale house was built only about 130 years ago. In the USA there are about a dozen houses nearing 100 years old that are still inhabited and showing no problems. They have an increasing stock of houses built since 1980 that are also surviving with no problems. Here in the UK, we started building in 1994, and 1996 in Ireland. As with any other technique of house building, if your straw bale house is built with a good design, with quality work and is properly maintained throughout its life, there is no reason why it should not last *at least* 100 years.

Isn't it a fire risk ?

No. It may seem strange, but when you stack bales up in a wall and plaster them either side, the density of the bales is such that there isn't enough air inside the bales for them to burn. It's like trying to burn a telephone directory—loose pages burn easily, but the whole book won't catch fire. Straw bale walls have passed all the fire tests they have been subjected to in the USA and Canada. Despite the bales themselves not being a risk, if you plaster any wall with a half inch of plaster, it gives sufficient fire protection to satisfy building regulations.

Is it really cheap to build ?

It depends entirely on your approach to building. If you can put lots of time into collecting recycled materials, or doing the drawings yourself and keep the design simple, or organise training workshops to build the walls and plaster them, or get your friends and family to help, then yes, it can be cheap to build. For most people, it is more sensible to think of doing the simple bits yourself (design, foundation, straw and plaster), and employing others to do the rest (carpentry, roofing, plumbing and electrics). A small 2-roomed building might cost about £10,000 (although substantially less with recycled materials), and a large 3-bedroomed house could be £40,000. Savings are greater on bigger buildings.

Can I do it myself ?

Yes, parts of it are quite easy to build. Other parts like roofing and carpentry are more difficult. It depends on how much time, determination and dedication you have. But the straw building technique is simple, straightforward and accessible to almost anyone.

What about temporary buildings ?

Design of straw bale buildings is very versatile, and can be adapted for a more or less durable function. If a building is only required for a few years, then there may be no need to build elaborate foundations, or plaster it inside or even outside.

What else can be built with straw ?

Straw has been put to many uses. Apart from houses, studios, offices and community spaces, straw is also used for schools, warehouses,

But the real point is that straw bale buildings are much cheaper to run once they've been built, because savings in energy/fuel costs due to the high insulation, can be as much as 75% on conventional building.

barns and stables, sound studios, meditation centres, acoustic barriers for airports and motorways, food storage and farm buildings.

What if some of my bales do get wet?

It depends on where, and how badly. Generally, if a bale gets wet through the top or bottom into the centre, then it will not dry out before it starts rotting. So any bales that are rained on, or stand in water whilst in storage, should be discarded. This also applies to any bales already in the walls that are not covered against the rain. But if you have covered the tops of the bales, and the sides get wet from the rain, this usually presents no problem, as they will quickly dry out once the rain stops. The only time this may not be the case is if the walls are exposed to severe wind and rain at the same time for prolonged periods, as the wind may drive the rain into the bale, where it cannot dry until the rain stops.

Is it possible to repair straw walls ?

It is not only possible, it's very easy! The hardest part is making a hole through the straw. This can be done with the claw on a hammer or crowbar, and by just pulling at the straw. It can be quite difficult to make the first hole due to the density of the bale. However, once this is done, wedges of the bales can be pulled out quite easily. Hazel pins can be cut through if necessary, and fresh straw wedges can be packed tightly back to fill the gap.

Experience has shown that if a section of wall does get wet, damp remains remarkably localised. It tends *not* to spread further through the straw, and so wedges or flakes of the bale can be removed and replaced.

What if I want an extra window?

Again, it's fairly easy to cut through the walls to create a window-sized hole. Usually, there is no need to support the rest of the wall as the wallplate carries most of the load, and the straw bales act together as an integral material because of the way they are pinned. Either follow the method above, or you can use a hayknife, even a chainsaw, although power tools like this tend to clog up very quickly. Once you've cut the hole, a structural boxframe can be fixed into the gap, with the window inside this.

Can I use straw to add an extension to my house?

Yes, both loadbearing and framed systems work well here. You may need to think carefully about settlement, and not make the final attachments from the straw to the house wall until after the walls are compressed.

You can also easily add an extension to your straw bale house by cutting a doorway through, in the same way as described above for making a window. Families have sometimes encouraged their children to build their own spaces once they've reached a suitable age!

Bibliography, References, Resources and Research

Books

The Beauty of Straw Bale Homes (2000) by Athena and Bill Steen. Chelsea Green Publishing Co. ISBN 1 890132 77 2. A wonderfully inspiring book showing just what it says; a range of pictures with brief descriptions of straw bale homes in the USA and Canada.

Build It with Bales Version Two (1997) by Matts Myhrman and S.O. Macdonald. Out On Bale Publishers. ISBN 0 9642821 1 9. This is the best and most 'hands on' manual for self-building with straw.

Building With Lime (1997) by Stafford Holmes and Michael Wingate. ITDG Publishing. ISBN 1 853393 84 3. An extensive handbook for construction uses of lime for floors, washes, wattle & daub, plasters, mouldings, mortars & more.

Buildings of Earth and Straw: Structural Design for Rammed Earth and Straw-Bale Architecture (1996) by Bruce King. Ecological Design Press. ISBN 0 964478 1 7. A technical book but written in an entirely accessible and entertaining way, for uninitiated builders and professionals alike, exploring the methods of building safe and durable straw and earth houses.

The Cob Builders Handbook (1997) by Becky Bee. Self-published. ISBN 0 9659082 0 8. Covers design, site selection, materials, foundations, floors, windows, doors, finishes & creative cob building techniques.

Earth Plasters for Straw Bale Homes (2000) by Keely Meagan. Self-published. Available through <www.strawbalecentral.com>. Covers earthen recipes, testing, problems, how to mix and apply each coat and tools.

The English Heritage Directory of Building Limes (1997) Donhead Publishing Ltd. ISBN 1 87339424 7. Manufacturers & suppliers in the UK & Ireland.

The Hand-Sculpted House: A Practical and Philosophical Guide to Building a Cob Cottage (2002) by Ianto Evans, Michael G. Smith and Linda Smiley. Chelsea Green Publishing Co (distributed in the UK by Green Books). ISBN 1 890132 34 9. Based on wide experience of cob construction.

Grand Designs (1999) by Kevin McCloud & Fanny Blake. Published by Channel 4 Books. ISBN 0 7522 1355 5. Building your dream home.

Serious Straw Bale: A Home Construction Guide for All Climates (2000) by Paul Lacinski & Michel Bergeron. Chelsea Green Publishing Co (distributed in the UK by Green Books). ISBN 1 890132 64 0. A Canadian book on design and build covering the serious issues of moisture, humidity and temperature.

Shelter (1973) by Lloyd Kahn. Shelter Publications Inc. ISBN 0 936070 11 0. A classic, fascinating book on the variety of structures possible.

Straw Bale Building: How to Plan, Design and Build with Straw (2000) by Chris Magwood and Peter Mack. New Society Publishers Ltd. ISBN 0 86571 403 7. A useful guide for the owner-builder.

Straw Bale Construction Details Book edited by Ken Haggard and Scott Clark. Available from strawbalecentral.com. Published by CASBA—good resource for designers, and owner-builders.

Strawbale Homebuilding (2000) Earth Garden Books. ISBN 0 9586397 4 4. Collection of Australian building experiences—but where they are still using cement plasters.

Straw Bale Details: A Manual for Designers and Builders by Chris Magwood and Chris Walker. New Society Publishers. ISBN 0 86571 476 2.

The Straw Bale House (1994) by Athena Swentzell Steen, Bill Steen & David Bainbridge. Chelsea Green Publishing Co (distributed in the UK by Green Books). ISBN 0 930031 71 7. An extremely popular and informative book based on the American experience, with beautiful, full colour photographs.

Straw for Fuel, Feed and Fertiliser (1982) by A.R. Staniforth. Farming Press Ltd. ISBN 0 85236 122 X. An interesting read about farming practice and the nature of straw.

Booklets and leaflets

* = available from <www.strawbalefutures.org.uk>
** = available from <www.strawbalecentral.com>
or <www.greenbuilder.com/dawn>

Appropriate Plasters for Cob and Stone Walls by the Devon Earth Building Association, c/o Environment Department, Devon County Council, County Hall, Exeter EX2 4QW. This pamphlet covers use of lime plasters and washes for protection and repair of cob and stone walls.

Compact Home Plans for Straw Bale and Superinsulated Construction by Community Ecodesign Network—Plans available to buy.**

The Green Building Handbook (1997) by Tom Woolley et al. Spon Press. ISBN 0419226907. A guide to building products and their impact on the environment.

A Guide to Straw Bale Building (1996, revised 2001) by Barbara Jones. Basic techniques of loadbearing construction, with information pack.*

House of Straw: Straw Bale Construction Comes of Age (1995) by US Department of Energy.**

How to Build with Straw Bales by Kevin Beale. Available from CAT Publications. <www.cat.org.uk>. A brief guide to straw bale building.

Information Guide to Straw Bale Building for Self-Builders & the Construction Industry (2001). By Barbara Jones. PDF Format.*

Lime in Buildings: A practical guide by Jane Schofield. Black Dog Press (1994). ISBN 0 9524342 1 3.

Using Lime by Bruce and Liz Induni. Self-published (1990). Available from the Society for Protection of Ancient Buildings, London.

A Visual Primer to Straw-Bale Construction in Mongolia by Steve MacDonald.**

Videos

Building with Straw Video Series by Black Range Films.**
Vol 1—A Straw Bale Workshop: View post and beam SB building at a weekend workshop.
Vol 2—A Straw Bale Home Tour: tour 10 homes ranging from low cost to luxury.
Vol 3—Straw Bale Code Testing: US building codes testing—impressive stuff.

How to Build your Elegant Home with Straw Bales by Sustainable Systems Support. Video and manual set for load bearing construction.**

Straw Bale Construction: Beautiful Sustainable Buildings by Straw House Herbals.**

Straw Bale Construction: The Elegant Solution by Sustainable Systems Support. Inspirational first video produced about straw bale construction in 1992.**

The Straw Bale Solution by NetWorks Productions. Overview of benefits of building with straw, featuring the work of Bill and Athena Steen in Mexico.**

* = available from <www.strawbalefutures.org.uk>
** = available from <www.strawbalecentral.com>
or <www.greenbuilder.com/dawn>

Resources

UK

Amazon Nails <www.strawbalefutures.org.uk>
<barbara@strawbalefutures.org.uk>

Association for Environment Conscious Building (AECB)
<www.aecb.net> <info@aecb.net>

The Building Limes Forum (BLF), Glasite Meeting House,
33 Barony Street, Edinburgh EH3 6NX
<admin@buildinglimesforum.org.uk>.

Centre for Alternative Technology (CAT) <info@cat.org.uk>
<www.cat.org.uk>

The Scottish Lime Centre <scotlime@aol.com>

Society for the Protection of Ancient Buildings (SPAB)
37 Spital Square, London E1 6DY. Tel: 0044 0207 3771644.
Has many leaflets on the use of lime.

**Straw Bale Building Association for Wales, Ireland, Scotland
& England** (WISE)
<www.strawbalebuildingassociation.org.uk>
<info@strawbalebuildingassociation.org.uk>

Women & Manual Trades <lwamt@dircon.co.uk>

Europe

European Straw Building Network (ESBN)
listserve: <strawbale-l@eyfa.org>

Austria: <www.baubiologie.at>

Belgium: <www.inti.be/ecotopie/ballots.html>

Chechnia: <www.fsv.cvut.cz/lists/ekodum/2001/msg00090.html>

Denmark: <www.folkecenter.dk/straw bale/inspirations-manual/inspirations-manual-1.html>

France: <www.la-maison-en-paille.com> &
<www.constructionfibres.citeweb.et/index.html>

listserve: <paille-l@eyfa.org>

CRATerre <www.craterre.archi.fr> <craterre@club-internet.fr>

Germany:

Hungary: <www.draconis.elte.hu/szalma/zemplen/zemplen.html>
 or text version <www.draconis.elte.hu/szalma>

Netherlands: <www.ndsm.nl/locatie/docs/houtenkop.html>
<www.rened.cistron.nl>

Norway: <www.strandsjo.no/htms/over-tysk.html>

Spain: listserve: <paja-l@eyfa.org>

USA & Canada

California Straw Building Association (CASBA)
<www.strawbuilding.org> <casba@strawbuilding.org>

The Canelo Project
<www.caneloproject.com> <absteen@dakotacom.net>

DAWN/Out on Bale by Mail
<www.greenbuilder.com/dawn> <dawnaz@earthlink.net>

Development Center for Appropriate Technology (DCAT)
<www.azstarnet.com/~dcat> <info@dcat.net>

The Last Straw <www.strawhomes.com>

Surfin' Straw Bale
<www.moxvox.com/surfsolo.html>
and <www.mha-net.org/html/sblinks.htm>

Straw Bale Central <www.strawbalecentral.com>

Australia

References and research

Bale Wall Compression Testing Programme
Lab-tested 2-string and 3-string walls at Colorado University in
1998. Info: <www.users.uswest.net/~jruppert2/odisea.htm>

Comparative Cost Analysis Between Building Methods
Investigates the economics of different construction techniques
Contact: Willow Whitton, 20819 NE Interlachen Ln, Troutdale, OR
97060, USA.

**Developing and Proof-Testing the 'Prestressed Nebraska'
Method for Improved Production of Baled Fibre Housing**
(1996) by Linda Chapman & Robert Plats. Test report documents
development and testing of a prestressed SB wall system.
Summary available from CMHC (Canadian Mortgage & Housing
Company) <www.cmhc-schl.gc.ca>

Evaluation of a Straw Bale Composite Wall (1999) by
Schmeckpeper & Allen. Tests performed on an unusual light-
gauge steel/straw bale wall system. Allen Engineering, 917 10th
Street, Clarkson, WA 99403, USA.

**Investigation of Environmental Impacts; Straw Bale
Construction** (1995) by Ann V. Edminster. In-depth investigation
of the environmental impacts of SB construction
<avedminster@earthlink.net>

Moisture in Straw Bale Housing: Nova Scotia (1998) by S.H.E
Consultants, Canada. <sheconsl@istar.ca>

**New Mexico ASTM E-119 Small Scale Fire Test & Structural
Testing** (1993) by Straw Bale Construction Association (SBCA) .
Includes SHB AGRA Lab report, Thermal Testing report from
Sandia National Lab and report from New Mexico Construction
Industries Division. Copies available through *The Last Straw*
Journal: <thelaststraw@strawhomes.com>

Pilot Study of Moisture Control in Stuccoed Straw Bale Walls
(1997). Illustrated report of the findings of a physical study into
the walls of several older Quebec-area SB structures to determine
how moisture is affecting them.
<www.cmhc-schl.gc.ca>

Straw Bale Construction Research Project by Portland
Community College Engineering Technology Dept. An ongoing
study of moisture levels in the walls of a small unoccupied
building in the Pacific Northwest, now in its fifth year. Joanna Karl
(<jkarl@pcc.edu>) or Lis Perlman (<lisp@iname.com>)

Straw Bales & Straw Bale Wall Systems (1993) by Ghailene Bou
Ali. Study of structural performance of bales and bale walls, which
influenced Tuscon building codes. Short illustrated report on this
research available as 'Summary of a Structural Straw-Bale Testing
Programme', available from TLS <thelaststraw@strawhomes.com>

Straw Bale Moisture Monitoring Report for the CMHC (1998).
Thorough reporting of four case studies in Alberta, Canada.
Summary available via email <robejoll@gyrd.ab.ca>

Structural Behaviour of Straw Bale Wall Construction (1998) by
John Carrick & John Glassford. Compressive, Transverse and
Racking load tests of 2-string rice straw bales as called up by the
Building Code of Australia.
<huffnpuff@shoal.net.au> <www.straw bale.archinet.com.au>

Testing Straw-Bale Construction in the Soggy North-West by
Aprovecho Research Centre <apro@efn.org>
<www.efn.org/~apro/straw bale.html>

Thermal & Mechanical Properties of Straw Bales as they Relate to a Straw House (1995) by K. Thompson, K. Watts, K. Wilkie and J. Corson. Reports on structural testing of bales and thermal & moisture monitoring of a SB house in Nova Scotia.
<kimt@chebucto.ns.ca>
<www.chebuto.ns.ca/~aa983strawhouse.html>

The Thermal Resistivity of Straw Bales for Construction (1993) by J C McCabe. Established R-values for wheat & rice straw bales.
<http://solstice.crest.rog/efficiency/straw_insulation/straw-insul.html>

All the above are available from <www.strawbalecentral.com> or <www.greenbuilder.com/dawn>

Also see *The Last Straw* issue no 31.

Appendix 3

Construction Drawings

Drawings on pages 107–115 are for Carymoor Environment Centre in Somerset, an example of a simple one-roomed dwelling.

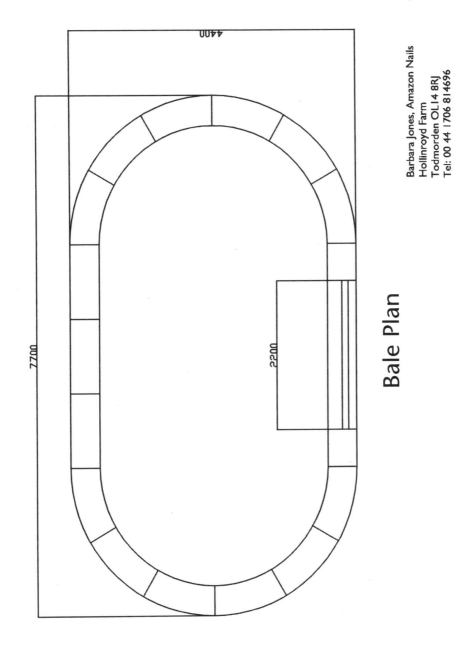

4400

7700

2200

Barbara Jones, Amazon Nails
Hollinroyd Farm
Todmorden OL14 8RJ
Tel: 00 44 1706 814696

Bale Plan

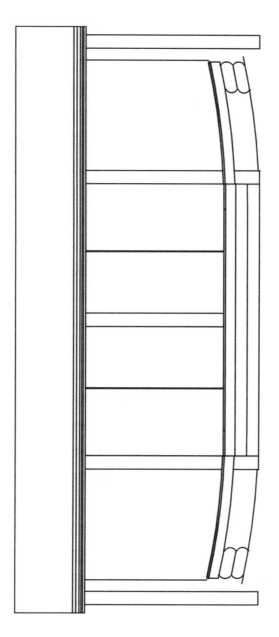

Front Elevation

Barbara Jones, Amazon Nails
Hollinroyd Farm
Todmorden OL14 8RJ
Tel: 00 44 1706 814696

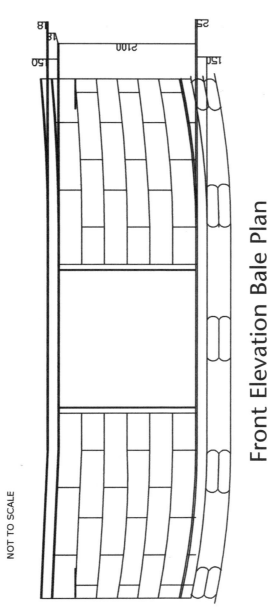

Front Elevation Bale Plan

NOT TO SCALE

Barbara Jones, Amazon Nails
Hollinroyd Farm
Todmorden OL14 8RJ
Tel: 00 44 1706 814696

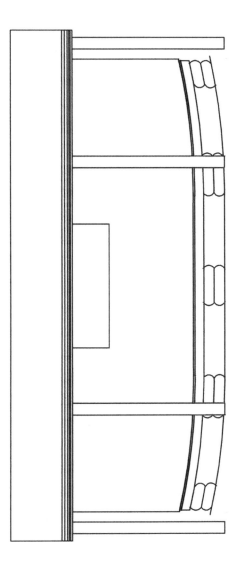

Rear Elevation

Barbara Jones, Amazon Nails
Hollinroyd Farm
Todmorden OL14 8RJ
Tel: 00 44 1706 814696

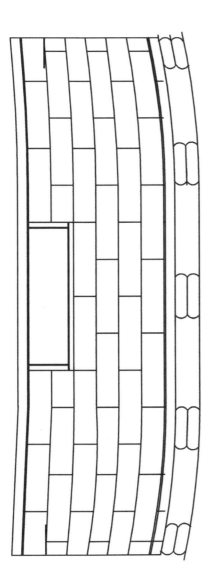

Rear Elevation Bale Plan

Barbara Jones, Amazon Nails
Hollinroyd Farm
Todmorden OL14 8RJ
Tel: 00 44 1706 814696

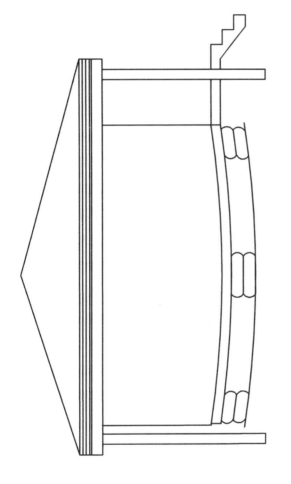

Side Elevation

Barbara Jones, Amazon Nails
Hollinroyd Farm
Todmorden OL14 8RJ
Tel: 00 44 1706 814696

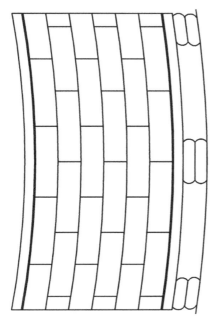

Side Elevation Bale Plan

Barbara Jones, Amazon Nails
Hollinroyd Farm
Todmorden OL14 8RJ
Tel: 00 44 1706 814696

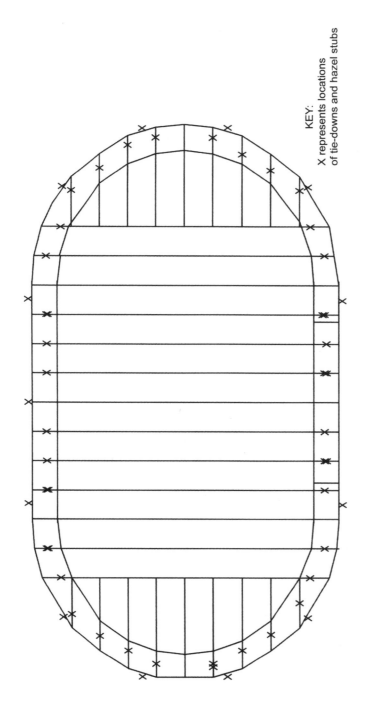

KEY:
X represents locations
of tie-downs and hazel stubs

Floor joists

Barbara Jones, Amazon Nails
Hollinroyd Farm
Todmorden OL14 8RJ
Tel: 00 44 1706 814696

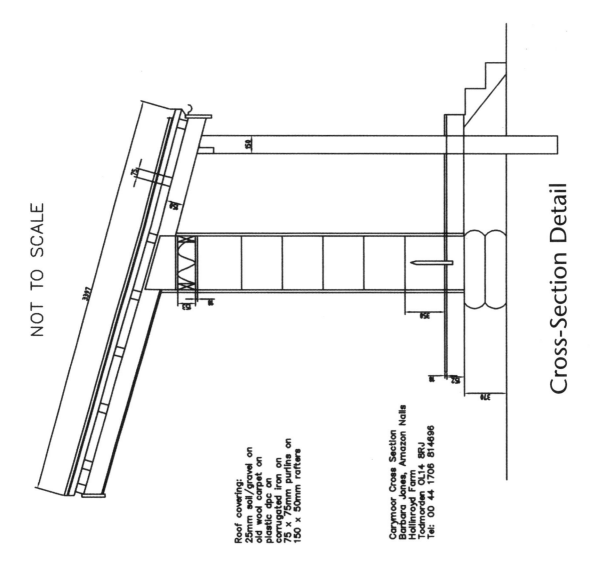

NOT TO SCALE

Roof covering:
25mm soil/gravel on
old wool carpet on
plastic dpc on
corrugated iron on
75 x 75mm purlins on
150 x 50mm rafters

Carymoor Cross Section
Barbara Jones, Amazon Nails
Hollinroyd Farm
Todmorden OL14 8RJ
Tel: 00 44 1706 814696

Cross-Section Detail

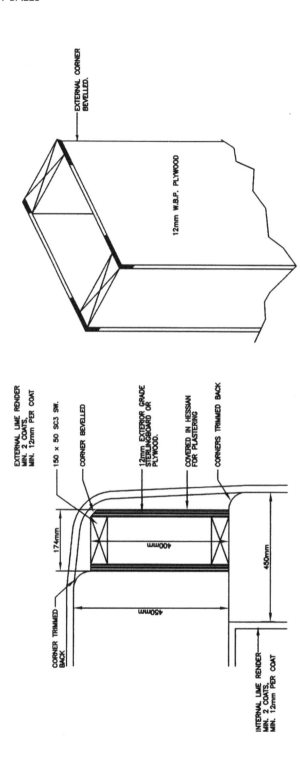

EXTERNAL CORNER BEVELLED.

12mm W.B.P. PLYWOOD

EXTERNAL LIME RENDER MIN. 2 COATS, MIN. 12mm PER COAT

150 x 50 SC3 SW.

CORNER BEVELLED

12mm EXTERIOR GRADE STERLINGBOARD OR PLYWOOD.

COVERED IN HESSIAN FOR PLASTERING

CORNERS TRIMMED BACK

174mm

400mm

450mm

450mm

CORNER TRIMMED BACK

INTERNAL LIME RENDER MIN. 2 COATS, MIN. 12mm PER COAT

Corner Post Detail

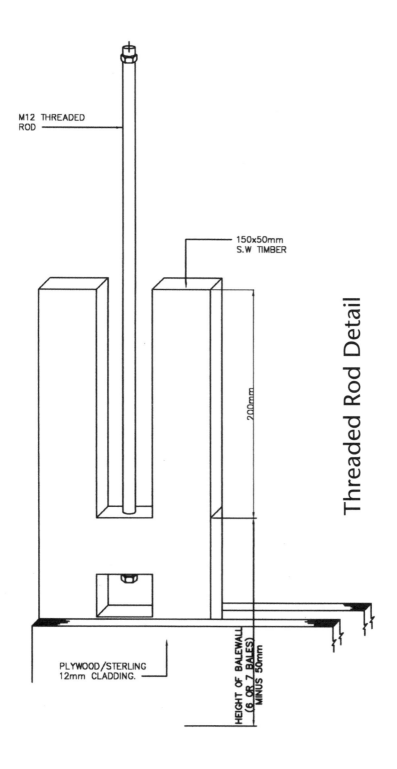

M12 THREADED
ROD

150x50mm
S.W TIMBER

200mm

Threaded Rod Detail

PLYWOOD/STERLING
12mm CLADDING.

HEIGHT OF BALEWALL
(6 OR 7 BALES)
MINUS 50mm

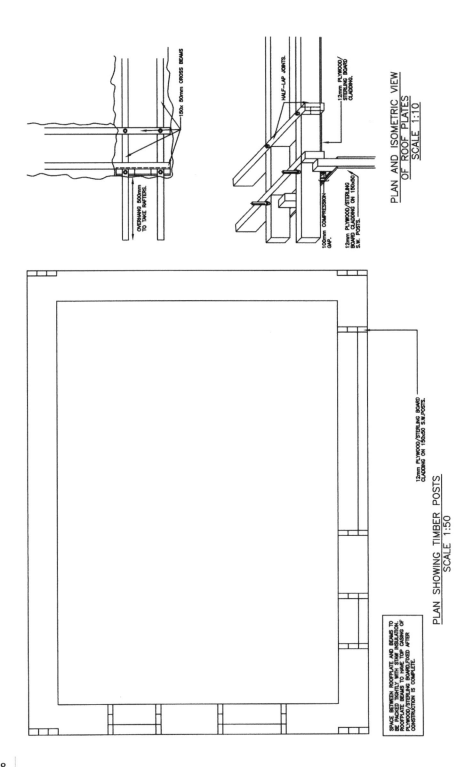

150x 50mm CROSS BEAMS

OVERHANG 500mm TO TAKE RAFTERS.

HALF–LAP JOINTS.

12mm PLYWOOD/STERLING BOARD CLADDING.

100mm COMPRESSION GAP.

12mm PLYWOOD/STERLING BOARD CLADDING ON 150x50 S.W. POSTS.

PLAN AND ISOMETRIC VIEW
OF ROOF PLATES
SCALE 1:10

12mm PLYWOOD/STERLING BOARD CLADDING ON 150x50 S.W.POSTS.

PLAN SHOWING TIMBER POSTS
SCALE 1:50

SPACE BETWEEN ROOFPLATE AND BEAMS TO BE PACKED TIGHTLY WITH STRAW INSULATION. ROOFPLATE AND BEAMS TO HAVE TOP CASING OF PLYWOOD/STERLING BOARD,FIXED AFTER CONSTRUCTION IS COMPLETE.

Post Layout and Wallplate

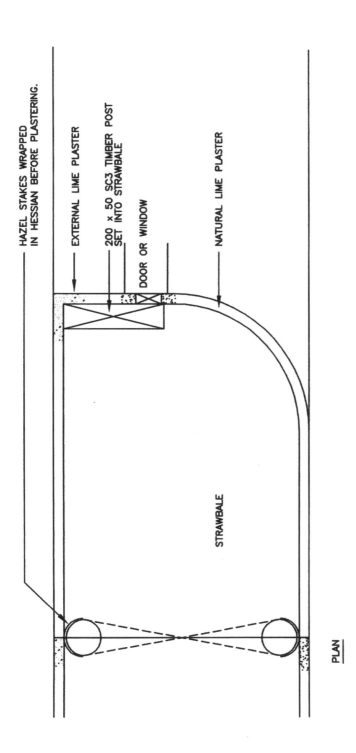

HAZEL STAKES WRAPPED
IN HESSIAN BEFORE PLASTERING.

EXTERNAL LIME PLASTER

200 x 50 SC3 TIMBER POST
SET INTO STRAWBALE

DOOR OR WINDOW

NATURAL LIME PLASTER

STRAWBALE

PLAN

Window and Door Reveals

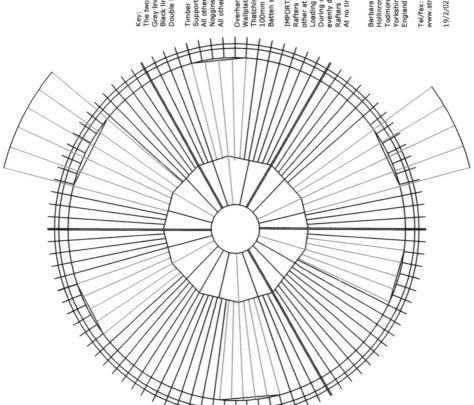

Rafters

Key:
The two outermost circles indicate location of wallplate
Grey lines indicate location of eyebrow and porch rafters
Black lines indicate location of common rafters
Double black lines indicate location of support rafters

Timber Dimensions:
Support rafters: 225 x 50mm
All other rafters 150 x 50mm
Noggins at 3m from chimney centre: 225 x 50mm
All other noggins: 150 x 50mm

Overhang of roof: minimum 500mm measured horizontally from bale face
Wallplate width: 350mm (50mm inset from bale faces)
Thatching battens: 50mm diameter hazel poles, securely nailed through to each rafter with
100mm galvanised round head nails
Batten spacings to be specified by thatcher

IMPORTANT
Rafters must sit on timber ledge at chimney and be securely fastened to ledge and to each
other at apex
Loading of roof at eaves level must be spread across whole width of wall
During construction, Rafters must be placed symmetrically around the roof to keep forces
evenly distributed around whole circumference of wall
Rafters must not be stacked together in one place on the roof
At no time shall an uneven weight be placed on any part of the roof

Barbara Jones, Amazon Nails
Hollinroyd Farm
Todmorden
Yorkshire OL14 8RJ
England

Tel/fax: 00 44 (0)1706 814696
www.strawbalefutures.org.uk

19/2/02

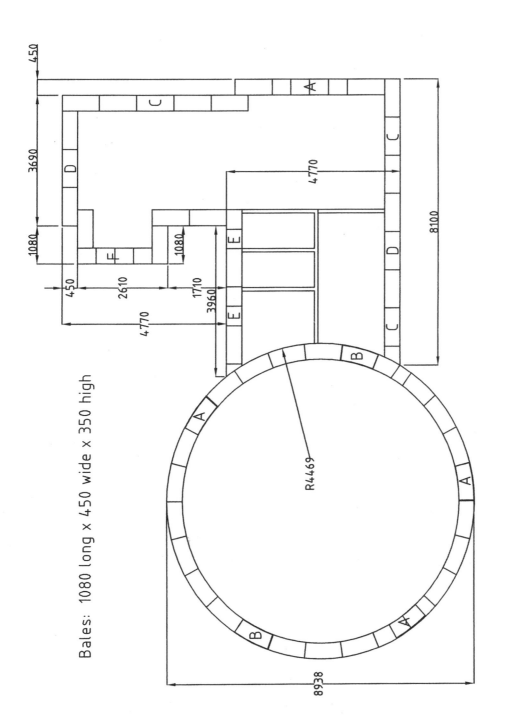

Bales: 1080 long x 450 wide x 350 high

Bale Plan for The Women's Centre, Co. Leitrim, Ireland

Index

Also available from Green Books & Chelsea Green

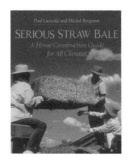

SERIOUS STRAW BALE A Home Construction Guide for All Climates Paul Lacinski and Michel Bergeron

Serious Straw Bale includes • thorough explanations of how moisture and temperature affect buildings in seasonal climates, with descriptions of the unique capacities of straw and other natural materials to provide warmth, quiet, and comfort year-round • comprehensive comparison of the two main approaches to straw bale construction: 'Nebraska-style', where bales bear the weight of the roof, and framed structures, where bales provide insulation • detailed advice including many well-considered cautions for contractors, owner-builders, and designers, following each stage of a bale-building process. Chelsea Green (USA) 480 pages with illustr. and b/w photos, 16pp colour section, resource section and index 252 x 204mm ISBN 1 890132 64 0 £22.50/$30.00 pb

THE STRAW BALE HOUSE

David Bainbridge, Athena & Bill Steen

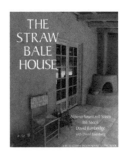

Shows how to build an entire house or something more modest—a studio, garage or children's playhouse—in 4-6 weeks for as little as half the cost of a conventional house; why straw is an excellent material for strength, energy- and resource-efficiency; how you can retro-fit inefficient or ageing buildings with straw bales, and gain high insulation values. Beautifully illustrated. Chelsea Green (USA) 250pp with 20 colour & 150 b/w photos, 45 illus. & plans 252 x 204mm ISBN 0 930031 71 7 £19.95/$30.00 pb

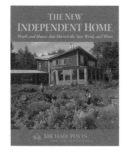

NEW INDEPENDENT HOME People and Houses that Harvest the Sun, Wind, and Water Michael Potts

Profiles the solar homesteaders whose innovations in building homes powered by sunlight, wind, and falling water have extended the possibilities of solar living. Beautiful houses are now being built with age-old materials, such as straw bales and rammed earth, and combined with state-of-the-art electronic technologies for harvesting free energy from the surrounding environment. Chelsea Green (USA) 350 pp 255 x 195mm 100 illustrations, 40 b/w photos & 16-page colour section ISBN 1 890132 14 4 £22.50/$30.00 pb

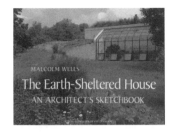

THE EARTH-SHELTERED HOUSE
An Architect's Sketchbook Malcolm Wells

Combine the thermal properties of earthen walls with simple principles of passive solar energy, and you have a foundation for unbeatable energy efficiency. Written in the author's own hand, and uses his futuristic illustrations. Chelsea Green (USA) 192 pp illus., b&w and col. photos 275 x 228mm ISBN 1 890132 19 5 £18.95/$24.95 pb

THE COB BUILDERS HANDBOOK Becky Bee

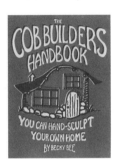

Cob (unbaked earth) is • environmentally friendly • comfortable: a house is warm in winter, cool in summer • quiet: cob walls muffle sound • long-lasting: houses can last for centuries. Has chapters on design, site selection, foundations, floors, windows and doors, roofs, finishes, and making and using cobs. Chelsea Green (USA) 172pp with 200 illustrations. 275 x 210mm ISBN 0 9659082 0 8 £17.95/$23.95 pb

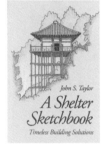

A SHELTER SKETCHBOOK John S. Taylor

Features • thousand year-old earth-sheltered houses in China • passive solar heating designs of the Pueblo Indians • 13th century Middle East air conditioning systems • modular building techniques used in Japan 500 years ago. Wonderful for stimulating the imagination! Chelsea Green (USA) 160pp with 550 illustr. 226 x 150mm ISBN 1 890132 02 0 £13.50/$18.95 pb

BUILDING WITH EARTH A Guide to
Flexible-Form Earthbag Construction Paulina Wojciechowska

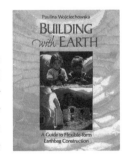

Describes the re-emergence of earthen architecture in North America, where adventurous builders are combining timeless forms such as arches, vaults and domes with modern materials and techniques. Using cheap recycled or salvaged polypropylene tubing or textile grain sacks, even relatively inexperienced builders can construct an essentially tree-free shelter, from foundation to curved roof, which can be used for retreats, studios and full-time homes in a variety of climates and conditions. Chelsea Green 176pp 252 x 202mm ISBN 1 890132 81 0 £19.95/$24.95 pb

THE NEW ECOLOGICAL HOME

A Complete Guide to Green Building Options

Daniel D. Chiras

Provides an overview of green building techniques, materials, products and technologies that are either currently available or will be in the near future. Includes chapters on: • The Healthy House • Green Building Materials • Wood-Wise Construction • Energy Efficiency • Earth-Sheltered Architecture • Passive Solar Heating and Passive Cooling • Green Power: Electricity from the Sun and Wind • Water and Waste • Environmental Landscaping. Chelsea Green (USA) ISBN 1 931498 16 4 352pp £25.00/$35.00 pb

THE HAND-SCULPTED HOUSE

A Practical and Philosophical Guide to Building a Cob Cottage

Ianto Evans, Michael G. Smith, and Linda Smiley

Theoretical and philosophical, but intensely practical as well: you will get all the information you need to undertake a cob building project. Cob houses, which literally rise up from the earth, are full of light, energy-efficient and cosy, with curved walls and built-in whimsical touches. Chelsea Green (USA) 384 pages with 8-page colour section and b/w illustr., appendices, source list, bibliog. and index ISBN 1 890132 34 9 £27.50/$35.00 pb

TO ORDER IN THE UK: phone Green Books on 01803 863260
sales@greenbooks.co.uk www.greenbooks.co.uk

TO ORDER IN THE USA: call Chelsea Green toll-free (1-800-639-4099)
or visit www.chelseagreen.com